Strategic Approaches to Problem Solving in Science and Engineering

This book serves as a comprehensive resource that highlights innovative practices and techniques within the realm of approximation methods. The book not only explores the theoretical aspects of approximation but also delves into its practical applications across a range of domains.

Features:

- Discusses the application of wavelets to develop computational schemes for solving mathematical models, offering a valuable approach to handle a variety of complex phenomena.
- Presents an inventory model designed for managing inevitable stock containing deteriorating objects, with a focus on time-dependent requirements and factors such as stationary holding and cost debilitation rates.
- Explores the numerical solution of two-dimensional, viscous magnetohydrodynamic (MHD) flow along a constant wedge through a porous medium.
- Showcases a stability analysis of thermal convection in a Jeffrey ferromagnetic fluid, focusing on both linear and weakly nonlinear aspects.
- Highlights an innovative mathematical modeling approach to address sustainability concerns in mixing problems, specifically considering the integration of renewable energy sources.

The text is primarily written for senior undergraduates, graduate students, and academic researchers in the fields including electrical engineering, electronics and communications engineering, computer science and engineering, manufacturing engineering, and industrial and production engineering.

Advancing Science and Engineering through Artificial Intelligence, Machine Learning, and Mathematical Modeling

Innovation in Healthtech: A Roadmap for Empowering Healthcare
Edited by: Rakesh Kumar and Meenu Gupta

Strategic Approaches to Problem Solving in Science and Engineering
Edited by Sharanjeet Dhawan, Ajender Kumar Malik, Manoj Sahni and Ritu Sahni

Strategic Approaches to Problem Solving in Science and Engineering

Edited by
Sharanjeet Dhawan
Ajender Kumar Malik
Manoj Sahni
Ritu Sahni

CRC Press
Taylor & Francis Group
Boca Raton London New York

CRC Press is an imprint of the
Taylor & Francis Group, an **informa** business

Designed cover image: © Shutterstock

First edition published 2027
by CRC Press
2385 NW Executive Center Drive, Suite 320, Boca Raton FL 33431

and by CRC Press
4 Park Square, Milton Park, Abingdon, Oxon, OX14 4RN

CRC Press is an imprint of Taylor & Francis Group, LLC

ISBN: 978-1-032-86110-4 (hbk)
ISBN: 978-1-032-86113-5 (pbk)
ISBN: 978-1-003-52138-9 (ebk)

DOI: 10.1201/9781003521389

Typeset in Sabon
by SPi Technologies India Pvt Ltd (Straive)

Contents

Preface

This book aims to provide valuable research guidance to the postgraduate students/research scholars/faculty members who are trying to make a mark in building their research work and to those seeking advancement in numerical methods. Here, we have discussed a collection of approximation methods with different fields such as fluid dynamics, nonlinear dynamics, inventory systems, fractional calculus, sustainability with renewable energy, machine learning, algorithms, fuzzy metric spaces, and integral transform. In Chapter 1, the authors discuss the stock-price dependencies for cost optimization in the presence of demand fluctuations, inflation, and backlogs. Chapter 2 deals with an inventory model with a learning effect under reliability-dependent demand. A stock model for deteriorating as well as debilitation items/objects, that is, the place rate of demand structured based on time, precipitously and stationary holding with cost debilitation rate, is explored in Chapter 3.

Chapter 4 presents the use of particle swarm optimization in an inventory model, which is a swarm-based intelligent search technique applied for solving nonlinear optimization problems in the inventory model. Carbon emissions and energy utilization have risen dramatically because of smart technologies, and they are directly related to different industries. Therefore, the renewable energy and environmental issues are receiving great attention across the globe, along with new technological developments for reducing global warming discussed in Chapter 5. Chapter 6 represents the linear and weakly nonlinear stability analysis of thermal convection in the Jeffrey ferromagnetic fluid, in which stability of the nonlinearity due to buoyancy magnetization is considered in the energy equation. A new approach for solving generalized functions in the fractional calculus is presented in Chapter 7. Chapter 8 is dedicated to the analysis of Quality of Services (QoS) for wireless sensor network. Chapter 9 discusses the design and development of the electromagnetic equipment with a proper analysis of pre applied currents, voltages, and power. The developed or amplified electrical quantities within the device, such as electric field, magnetic field, electric potential, electron velocities and energies, and space charge forces, must be carried out to depict the overall performance of the device design.

Analysis of AI-Based Emotion Recognition Techniques in Online Learning Systems is done in Chapter 10. The Chapter 11 deals with the approximation in the dynamics of chaotic maps using modified Mann orbit. In Chapter 12, authors discuss the physics informed neural networks (PINNs) to analyze real-world transport processes. Efficiency of the meshfree PINN framework is highlighted over the traditional techniques. In Chapter 13, the authors discuss content market performance measurement through the OWA operator.

About the editors

Sharanjeet Dhawan, an assistant professor of Mathematics at CCSHAU, Haryana, India, and a core committee member in Mathematical Sciences for the Indian National Young Academy of Sciences (INYAS), conducts research that bridges theoretical and applied mathematics to address real-world challenges through advanced computational techniques. Her work focuses on developing and implementing wavelet collocation methods, finite element analysis, and numerical simulations for solving differential equations arising in fluid mechanics, biomechanics, and nonlinear wave dynamics, resulting in numerous high-impact publications. Through extensive international collaborations, she has significantly advanced computational methodologies for partial differential equations while actively promoting gender equity in science as a member of the Organization for Women in Science for the Developing World (OWSD) and a founding member of the Women in STEM Network. Dr. Dhawan has organized and participated in numerous national and international conferences and maintains active life membership in several learned societies, including OWSD, the Ramanujan Mathematical Society, and the Indian Mathematical Society, with her primary research interests centered on differential equations and numerical analysis.

Ajender Kumar Malik, Professor of Mathematics and Director, Centre for Online Education at UP Rajarshi Tendon Open University, Prayagraj, Uttar Pradesh, India. He is also Hon. (Voluntary) National Director of the National Council of Researcher Teacher Scientists India, affiliated to VIPNET, Vigyan Prasar (under DST Govt. of India). Recently, he has been honored a Shikshak Samman: National Teacher Award 2022 on February 1, 2022 and also honored a Srinivasa Ramanujan Award for Mathematics Popularization on December 22, 2018, on National Mathematics Day from VIPNET-Vigyan Prasar (under DST Govt. of India) Network DASA & SWA of Tripura. He has published more than 100 research papers/book chapters, delivered more than 100 talks/paper presentations, and worked as a visiting Professor in EISTI University,

CERGY, Paris, France. He has supervised 10 PhD students and is author/co-author of 31 books like *Applications of Advanced Optimization Techniques in Industrial Engineering* (CRC Press), *Operations Research, Optimization Techniques, Topology, Measure Theory & Integration* etc.

Manoj Sahni is a dedicated and experienced Mathematics teacher and researcher with more than 20 years of experience and currently serving as a professor in the Department of Mathematics, School of Technology, Pandit Deendayal Energy University, Gandhinagar, Gujarat, India. He has an excellent academic background with an MSc (Mathematics with specialization in Computer Applications) from Dayalbagh Educational Institute (Deemed University), Agra, MPhil from IIT Roorkee, and a PhD in Mathematics from Jaypee Institute of Information Technology (Deemed University), Noida, India. He has published more than 120 research papers in peer-reviewed journals (SCI/SCIE/ESCI/Scopus), conference proceedings (Scopus Indexed), and book chapters (Scopus Indexed) with reputed publishers. He also serves as an advisory board member, technical committee member, and reviewer for many reputed international journals and conferences. He has organized five International Conferences on Mathematical Modeling, Computational Intelligence Techniques, and Renewable Energy (MMCITRE) from 2020 to 2025. He has published 12 edited books, and three more are in the pipeline. He has received two national awards and was nominated by the Gujarat Government for the National Teacher's award. He participated in the scientific committee of several international conferences and associations and also delivered many expert talks at the national and international levels. He has organized many seminars, workshops, short-term training programs at PDEU, and various other universities. In addition, he is a member of many international professional societies, including the American Mathematical Society (AMS), Society for Industrial and Applied Mathematics (SIAM), IEEE, Mathematical Association of America (MAA), Forum for Interdisciplinary Mathematics (FIM), Indian Mathematical Society (IMS), IAENG, and many more. He has guided five UG, one PG, one JRF, and four PhD students, and three more are working under him.

Ritu Sahni is a dedicated and experienced Mathematics teacher and researcher with more than 18 years of experience who is actively involved in teaching both undergraduate and postgraduate levels. She has a strong academic background having completed BSc (Mathematics), MSc (Mathematics with specialization in Computer Applications) from Dayalbagh Educational Institute, Agra, UP, India, and PhD from Jaypee Institute of Information Technology (JIIT), Noida, UP, India. During her PhD, she also served her institution (JIIT, Noida, India) by teaching undergraduate and postgraduate students and after the completion of her Doctorate, she served various institutions including Navrachna University, Vadodara, Gujarat, India, Institute of Advanced Research

(IAR), Gandhinagar, India as an assistant professor, as Visiting Faculty at various universities including Pandit Deendayal Energy University, Gandhinagar, Adani Institute of Infrastructure, Ahmedabad, and Kaushalya University, Ahmedabad. Presently, she is working as a visiting faculty at Nirma University, Ahmedabad. She worked with her students on various research-related problems and published their work in reputed journals (SCI/SCIE/ESCI/Scopus). She has published more than 40 research papers in peer-reviewed international journals and conferences. She is the reviewer of many international journals and is a member of the Indian Science Congress and many other well-renowned societies. She has also organized seminars, workshops, and coordinated the 2nd and 3rd International Conferences on Mathematical Modelling, Computational Intelligence Techniques and Renewable Energy (MMCITRE) held from February 6–8, 2021 and March 4–6, 2022, respectively. She has delivered talks and presented many research articles in various international conferences in India and abroad. Her research field includes fixed point theory and its applications, numerical methods, fuzzy decision-making systems, and other allied areas. She is a devoted faculty member, committed researcher, and enthusiastic about striving to improve the institution's educational offerings and research fields for societal growth.

Contributors

Ashish
Department of Mathematics
Govt. College Satnali
Mahendergarh, India

Dinesh Bhatia
Department +of Textile Engineering
Jawaharlal Nehru Government Engineering College (JNGEC)
Sundernagar, India

Jinde Cao
Dean, School of Mathematics,
Director of the Jiangsu Provincial Key Laboratory of Networked Collective Intelligence of China
China

Alicia Delgadillo-Aguirre
Universidad Autónoma de Occidente
Culiacán, México

Sharanjeet Dhawan
Department of Mathematics
CCSHAU COA Bawal
Hisar, India

Vishant Gahlaut
Department of Physics
Banasthali Vidyapith
Jaipur, India

Abhinav Goel
Department of Mathematics at Graphic Era
Deemed to be University
Dehradun, India

Vinti Gupta
Department of Computer Science and Engineering
ABES Engineering College
Ghaziabad, India

Martín Huesca-Gastélum
Universidad Tecnológica de Culiacán
Culiacán Rosales, México

Mahesh Kumar Jayaswal
Department of Applied Science and Humanities
School of Engineering and Sciences
Art, Design and Technology University
Pune, India

Meenu Kaushik
School of Automation
Banasthali Vidyapith
Rajasthan, India

Kavita
Department of Mathematics
Vallabh Government College
Mandi, India

Faizan Ahmad Khan
Department of Mathematics
Faculty of Science
University of Tabuk
Tabuk, Saudi Arabia

Ernesto León-Castro
Universidad Católica de la Santísima Concepción
Concepción, Chile

Ajender Kumar Malik
Centre for Online Education,
UP Rajarshi Tandon Open University
Prayagraj, India

P N Pathak
Department of Mathematics
School of Basic Sciences CSJM University
Kanpur, U.P., India

Luis Pérez-Arellano
Universidad Autónoma de Sinaloa
Culiacán Rosales, México

Ather Aziz Raina
Department of Applied Mathematics
Government Degree College Thannamandi
Jammu & Kashmir, India

Pooja Rani
Department of Mathematics & Statistics
COBS&H, CCSHAU
Hisar, India

Atul Sahu
School of Sciences,
Uttar Pradesh Rajarshi Tandon Open University (UPRTOU)
Prayagraj, India

Amit Saini
Department of Mathematics
Dewan Institute of Management and Studies
Meerut, India

Animesh Kumar Sharma
Faculty of Science and technology
The ICFAI University Raipur
Kumhari, India

Archana Sharma
Department of Applied Sciences
KIET Deemed to be University
Ghaziabad, India

Veena Sharma
Department of Mathematics
Himachal Pradesh University (HPU)
Shimla, India

Chandrakant Kumar Singh
School of Sciences
Uttar Pradesh Rajarshi Tandon Open University (UPRTOU)
Prayagraj, India

Dipti Singh
Department of Mathematics
Chaudhary Charan Singh University (CCS University)
Meerut, India

Raghvendra Singh
School of Sciences
Uttar Pradesh Rajarshi Tandon Open University
Prayagraj, India

Rajendra Singh
Department of Commerce
Dayanand Anglo-Vedic (PG) College Civil Lines
Kanpur, India

Shiv Raj Singh
Department of Mathematics
Chaudhary Charan Singh University (CCS University)
Meerut, India

Vishal Singh
School of Interdisciplinary Design and Innovation,
Indian Institute of Information Technology Design and Manufacturing
IIITDM Kancheepuram, India

Claudia Tirado-Gálvez
Universidad Autónoma de Sinaloa
Culiacán Rosales, México.

Vinam Tomar
Department of Computer Science and Engineering
SRM Institute of Science and Technology
Delhi, India

Priyanka Verma
Department of Mathematics
Bhagwant University
Ajmer, India

Sachin Kumar Verma
Department of Mathematics
Dewan Institute of Management and Technology
Meerut, India

Chapter 1

Exploring stock-price dependencies for cost optimization in the presence of demand fluctuations, inflation, and backlogs

Sachin Kumar Verma, Amit Saini, Archana Sharma, and Ajender Kumar Malik

1.1 INTRODUCTION

The demand for a product or service represents the quantity consumers are prepared to purchase at different price levels. This demand is influenced by several key factors, including price, consumer preferences, income levels, and the presence of substitute options. The law of demand states that as prices increase, the quantity demanded generally decreases, and conversely, as prices decrease, the quantity demanded typically rises. A number of variables, including timing, price, and supply levels, affect demand. The link between the amount desired and the price of a product is known as price-dependent demand; as the price rises, demand falls. The quantity of accessible stock has an impact on stock-dependent demand; whereas lower stock levels reduce demand, larger stock levels increase it due to improved visibility or the feeling of plenty among consumers. Since these elements have the potential to greatly affect a product's total demand, it is essential to comprehend how they interact to anticipate demand effectively.

Abad (1996) discussed optimal order policies for pricing and stock size situations under the conditions of a half-done backorder inventory system. Goyal et al. (1996) explored a model in which a special kind of trend is followed in this inventory model. In this model, the size of the stock-linked problem has been discussed with scarcities under a linear trend-based demand pattern. Jaggi and Khanna (2010) invented a model of supply chain control system for deteriorating items with a stock-connected ingesting rate, which is connected with the shortages and inflation; in this model, the buyer and seller exchange of payment was the key focus point. Lee and Dye (2012) generated a model of inventory items for in-hand inventory dependent demand connected with the manageable deterioration rate. Padmanabhan and Virat (1995) discussed an economic order quantity (EOQ) model for unpreserved items with stock- and selling price-based demand rate. Rathore et al. (2024) addressed the increasing demand over time while controlling costs by introducing an inventory strategy with ramp-type demand and

DOI: 10.1201/9781003521389-1

time-varying holding costs in a two-warehouse system. The model is developed to efficiently manage costs and satisfy demand spikes in a dual-storage setting.

A model for non-instantaneously degrading products with price- and stock-sensitive demand was also created by Rezagholifam et al. (2022), with an emphasis on capacity-constrained optimum pricing and ordering. By matching demand sensitivity with ordering choices, this approach assists businesses in meeting capacity constraints and controlling pricing strategies for degrading inventory for two stores under a single management system with inflation. Singh and Sharma (2014) presented a policy for a model in which trade and credit between the buyers and suppliers have been the key point with the stock-based demand for perishable items. Singh et al. (2010a) developed a model based on deteriorating items in the case of shortages. The main focus was on the available stock-dependent demand under the effect of inflation for a couple of shops under the one system. Singh et al. (2010b) formulated a model in which a policy for replenishment has been made for the product, which is non-instantaneous in nature. The model is connected with the two stores under the case of inflation. Additionally, Thilagavathi et al. (2022) created a two-warehouse inventory model with partial backlog and stock-dependent demand that incorporates a three-slot payment policy under inflation. This concept aids businesses in cash flow management. This concept aids businesses in cash flow management and ideal ordering strategy with stock-dependent demand that takes into consideration acceptable payment delays that were put out by Tripathi and Mishra (2013). By modifying ordering schedules to take advantage of longer payment terms, this technique helps manage demand-sensitive inventory and improves cash flow in businesses with fluctuating demand. The Goel et al. (2015) model addresses a supply chain where demand increases with stock levels (stock-dependent demand) and items deteriorate at an accelerating, quadratic rate. It also allows for temporary shortages, which are managed to meet demand later. This model is especially useful for optimizing inventory in cases involving perishable goods, balancing demand, deterioration costs, and shortage handling.

Palanivel et al. (2016) proposed a two-warehouse inventory model for items that deteriorate gradually, with demand increasing as stock levels rise. The model includes shortages, managed by backordering, and accounts for inflation to better estimate inventory costs. This approach helps optimize storage across two warehouses by considering deterioration, stock-sensitive demand, and economic fluctuations.

Aarya and Kumar (2018) presented a model reflecting customer sensitivity toward stock- and price-linked demand. The model allows for partial backlogging, meaning a portion of unmet demand is backordered and fulfilled later. This approach helps balance production rates, inventory costs, and customer satisfaction by optimizing inventory levels according to price, stock, and demand.

Hsieh et al. (2010) presented a model to determine the optimal lot size for items where demand increases when stock exceeds a specific threshold, a condition known as stock-dependent demand. The model includes partial backlogging, allowing a portion of unmet demand to be fulfilled later. This setup helps businesses decide optimal order quantities by balancing high-demand periods, inventory levels, and controlled shortages.

Generally, the shortage case was avoided by researchers in the initial stage of research. It was found that in real-life situations, a shortage occurs at certain points in time. So, shortages cannot be avoided at the time when discussing the case of inventory models of deterioration items. Abad (2001) presented optimization of total cost in which price is taken as the key element for making a policy for ordering. This model is based on the size of the lot and reselling after selling it once already under shortage, considering customer return and departure situation. Dye (2007) introduced the concept of joint pricing with the ordering policy for deteriorated nature inventory with price-sensitive demand and customer return situations. Zhou et al. (2004) discussed the issue of waiting time based on the partial backlogging rate with time-dependent demand.

Price hike or decrease depend on the various factors. Out of which, inflation is the key component in managing the inventory system, generally researchers avoid the effect of inflation in courting total cost. But in real-life situations, inflation cannot be ignored. So, in the past couple of years, researchers focused on inflation in developing various models of deteriorating items. Buzacott (1975) developed a model in which economic order quantities under with an inflation-based problem has been discussed. Jain and Singh (2011) developed a model considering inflationary consequences on inventory in the situation when the date connected with capital return or payment has expired under the unpredictable lead time in the system of a multi-echelon supply chain. Kumar et al. (2012) emphasized the supply chain of three echelon kind of model with limited available storage capacity and lead time under an inflationary environment. Ouyang et al. (2003) discussed a model for deteriorating items with stock-dependent demand under inflation and return value of money over time. Sarkar et al. (2010) explored a finite case of replacement of the inventory model with increasing demand under inflation in real-life situation case.

Sarkar et al. (2011) worked on a model in which a deficient production process for time-varying demand with inflation related to the time value of money has been discussed. Singh and Singh (2011) considered a model that is not perfect for the production process with an exponential demand-based rate connected with a Weibull deterioration rate under the effect of inflation. Udayakumar et al. (2020) suggested a dynamic economic ordering strategy that includes acceptable payment delays under inflation. Both Yang and Chang (2013) and Yang (2006) provided models for two-warehouse systems with partial backlogs for degrading products. These models included payment deferrals and inflationary pressures. The requirement for adaptable inventory systems that are sensitive to inflation and can adjust to the ever-changing economic environments of today is highlighted by these models taken together.

In this chapter, the impact of customer returns under an inflationary environment and partial backlogging on the inventory system of deteriorating items has been discussed. The multivariate demand depending on stock, time, and price has been considered. To the best of our knowledge, no researcher has worked on the concept of customer returns with two different demands in a shortage and non-shortage periods under inflation. In the shortage phase, time- and price-dependent demand rate has been assumed.

This chapter systematically presents its research, starting with assumptions with notations in Section 1.2, followed by the development of a multivariate demand model in Section 1.3. Section 1.4 outlines the problem-solving process, while Sections 1.5 and 1.6 provide a numerical example and sensitivity analysis to demonstrate the model's practical use, respectively. Finally, results analysis, managerial implications, and conclusion highlight the key findings and effectiveness of the model in Sections 1.6.1, 1.6.2, and 1.7, respectively.

1.2 ASSUMPTIONS AND NOTATIONS

1.2.1 Assumptions

This model was developed using the following presumptions.

1. $D_1(p, I(t)) = (a + \alpha I(t))p^{-\beta}$ and $D_2(p) = ap^{-\beta}$ are the demand function stock-in and stock-out situations, respectively, where α is the positive constant, $\alpha < 1$, and β is the positive constant, $\beta > 1$
2. It is acceptable to have shortages and to partially backlog at an exponentially diminishing pace, as $B(T-t) = e^{-\delta(T-t)}$ where $\delta > 0$.
3. The deteriorated items are completely discarded.
4. Inventory items are deteriorating in nature.
5. Inflation is allowed.

1.2.2 Notations

The following notations are used in the development of this model.

A	Ordering cost	Per order
C_p	Purchasing cost per unit	Per unit
C_h	Holding cost	
C_s	Shortage cost	
C_l	Lost sale cost	
p	Selling price	

(Continued)

θ	Deterioration rate	Constant
R	Inflation	
q_1	Stock level at $t = 0$	
q_2	Stock level at $t = T$	
Q	Order quantity	
T	Replenishment cycle length	Time unit
t_1	Stock in time	

1.3 MATHEMATICAL FORMULATION

A lot size of q_1 units is received at $t = 0$ at the start of the inventory model cycle (Figure 1.1). It starts depleting due to demand and deterioration and drops to zero at $t = t_1$. After that, shortages accumulate during the time interval $[t_1, T]$ and are partially backlogged. At $t = T$, the level of negative inventory reaches q_2 units. The equation depicting the inventory level across several time periods is provided as follows:

$$\frac{dI_1(t)}{dt} = -\theta I_1(t) - \frac{(a + \alpha I_1(t))}{p^\beta} \quad 0 \le t \le t_1 \tag{1.1}$$

$$\frac{dI_2(t)}{dt} = -B(T - t)\frac{a}{p^\beta} \quad t_1 \le t \le T \tag{1.2}$$

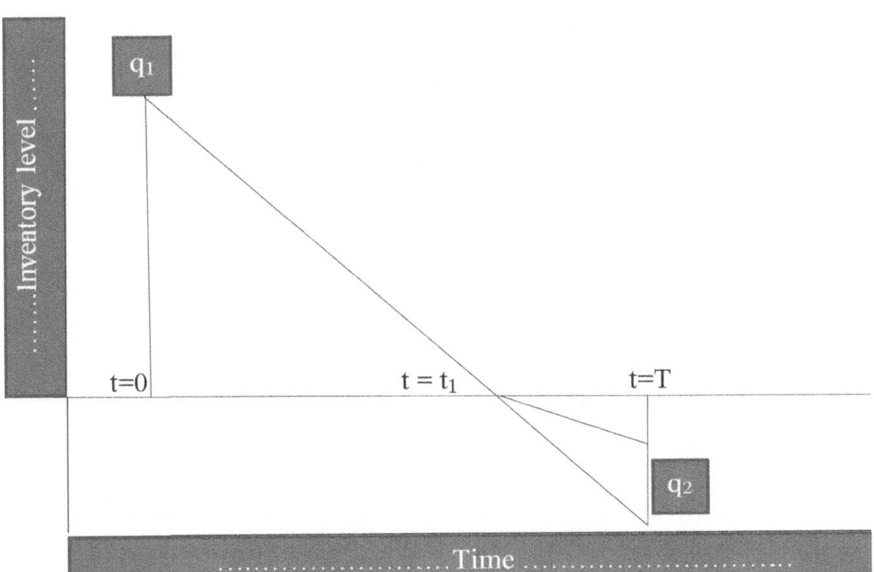

Figure 1.1 Graphical representation of inventory with time.

With boundary conditions, $I_1(t_1) = I_2(t_1) = 0$, the following are solutions of (1.1) and (1.2):

$$I_1(t) = \frac{a}{(\alpha + p^\beta \theta)} \left(e^{\left(\frac{\alpha}{p^\beta} + \theta\right)(t_1 - t)} - 1 \right) \quad (1.3)$$

$$I_2(t) = \frac{a}{\delta p^\beta} \left(e^{-\delta(T - t_1)} - e^{-\delta(T - t)} \right) \quad (1.4)$$

The maximum positive inventory is

$$q_1 = I_1(0) = \frac{a}{(\alpha + p^\beta \theta)} \left(e^{\left(\frac{\alpha}{p^\beta} + \theta\right) t_1} - 1 \right) \quad (1.5)$$

The maximum backorders are

$$q_2 = -I_2(T) = \frac{a}{\delta p^\beta} \left[1 - e^{-\delta(T - t_1)} \right] \quad (1.6)$$

Total ordering quantity

$$Q = [q_1 + q_2] = \left[\frac{a}{(\alpha + p^\beta \theta)} \left(e^{\left(\frac{\alpha}{p^\beta} + \theta\right) t_1} - 1 \right) + \frac{a}{\delta p^\beta} \left[1 - e^{-\delta(T - t_1)} \right] \right] \quad (1.7)$$

The cost per cycle consists of the following costs:

1. **Ordering cost**

$$OC = A \quad (1.8)$$

2. **Holding cost**

$$HC = C_h \int_0^{t_1} I_1(t) e^{-Rt} dt$$

$$= C_h \left[\frac{1}{(\alpha + p^\beta \theta)} \left(\frac{a e^{-Rt_1} \left(e^{t_1 (R + p^{-\beta} \alpha + \theta)} - 1 \right)}{(\alpha p^{-\beta} + (R + \theta))} - \frac{a(1 - e^{-Rt_1})}{R} \right) \right] \quad (1.9)$$

3. **Shortage cost**

$$SC = C_s \int_{t_1}^{T} \left[-I_2(t) \right] e^{-Rt} dt$$
$$= C_s \left[\frac{a e^{-R(T+t_1)-T\delta} p^{-\beta} \left(e^{Rt_1+T\delta} R - e^{RT+t_1\delta} \delta + e^{t_1(R+\delta)} (-R+\delta) \right)}{R(R-\delta)\delta} \right] \tag{1.10}$$

4. **Lost Sale cost**

$$LSC = C_l \int_{v}^{T} \left(1 - B(T-t)\right) D_2(p) e^{-Rt} dt$$
$$= C_l \int_{v}^{T} \left(1 - e^{-\delta(T-t)}\right) \frac{a}{p^\beta} e^{-Rt} dt \tag{1.11}$$
$$= C_l \left[\frac{a}{p^\beta} \frac{e^{-R(T+t_1)} \left(e^{Rt_1} \delta - e^{RT} (-R+\delta) - e^{RT+(-T+t_1)\delta} R \right)}{R(R-\delta)} \right]$$

5. **Purchasing cost**

$$PC = C_p [q_1 + q_2] e^{-RT}$$
$$= C_p \left[\frac{a}{(\alpha + p^\beta \theta)} \left(e^{\left(\frac{\alpha}{p^\beta} + \theta\right) t_1} - 1 \right) + \frac{a}{\delta p^\beta} \left[1 - e^{-\delta(T-t_1)} \right] \right] e^{-RT}$$

6. **The present worth total cost of the system is**

$$TC(t_1, p) = \frac{1}{T} [OC + HC + BC + LSC + PC] \tag{1.12}$$

1.4 SOLUTION METHOD

Step 1: Find $\frac{\partial TC}{\partial t_1}, \frac{\partial TC}{\partial p}, \left(\frac{\partial^2 TC}{\partial {t_1}^2}\right), \left(\frac{\partial^2 TC}{\partial p^2}\right)$, and $\left(\frac{\partial^2 TC}{\partial t_1 \partial p}\right)$.

Step 2: $\frac{\partial TC}{\partial t_1} = 0$ and $\frac{\partial TC}{\partial p} = 0$. after solving this, assume that the values that meet the above equations are t_1^* and p^*.

Step 3: The following requirements will be met if the goal function is concave.

$$\text{If } \left(\frac{\partial^2 TC}{\partial t_1^{\ 2}}\right)_{(t_1^*,p^*)}\left(\frac{\partial^2 TC}{\partial p^2}\right)_{(t_1^*,p^*)} - \left(\frac{\partial^2 TC}{\partial t_1 \partial p}\right)_{(t_1^*,p^*)} > 0 \text{ and } \left(\frac{\partial^2 TC}{\partial t_1^{\ 2}}\right)_{(t_1^*,p^*)} > 0$$

1.5 NUMERICAL EXAMPLE

The following parameters are used in the appropriate units: $A = 500, a = 30,$ $\alpha = 0.5, \beta = 0.6, R = 0.03, \delta = 0.7, C_s = 15$ per unit, $C_l = 10$ per unit, $\theta = 0.2,$ $C_b = 15$ per unit, $C_p = 10$ per unit, $T = 12$ Unit time. The decision variable's ideal values are

$$t_1^* = 6.5843 p^* = 17.392 Q^* = 303.306, \text{ and } TC^* = 1326.76.$$

1.6 SENSITIVITY ANALYSIS

Effect of different parameters on optimal policy

Parameters	*% Change*	*Value*	*p*	*t₁*	*TC*
R	50	0.045	17.392	6.2217	1156.61
	25	0.0375	17.392	6.4112	1242.63
	0	0.03	17.392	6.5843	1326.76
	−25	0.0225	17.392	6.7421	1408.82
	−50	0.0015	17.392	7.1122	1631.45
δ	50	1.05	17.392	7.0335	1400.34
	25	0.875	17.392	6.8194	1365.19
	0	0.7	17.392	6.5843	1326.76
	−25	0.525	17.392	6.3247	1284.08
	−50	0.35	17.392	6.0363	1236.91
C_s	50	22.5	17.392	7.3930	1462.62
	25	18.75	17.392	7.0212	1400.30
	0	15	17.392	6.5843	1326.76
	−25	11.25	17.392	6.0637	1238.15
	−50	7.5	17.392	5.4328	1129.99
C_l	50	15	17.392	7.0335	1400.34
	25	12.5	17.392	6.8194	1365.19
	0	10	17.392	6.5843	1326.76
	−25	7.5	17.392	6.3247	1284.08
	−50	5	17.392	6.0363	1236.91

(Continued)

C_h	50	22.5	17.392	5.4056	1590.81
	25	18.75	17.392	5.9356	1471.29
	0	15	17.392	6.5843	1326.76
	−25	11.25	17.392	7.3986	1209.53
	−50	7.5	17.392	8.4554	920.29
T	50	18	17.392	9.3326	1583.92
	25	15	17.392	8.0140	1499.79
	0	12	17.392	6.5843	1326.76
	−25	9	17.392	5.0565	1127.05
	−50	6	17.392	3.4428	910.87
a	50	45	17.492	9.3526	910.007
	25	37.5	17.490	8.0440	1117.05
	0	30	17.392	6.5843	1326.76
	−25	22.5	17.391	5.0665	1489.79
	−50	15	17.390	3.4428	1593.92
α	50	0.75	17.992	9.3336	1083.92
	25	0.625	17.892	9.0140	1199.79
	0	0.5	17.392	6.5843	1326.76
	−25	0.375	17.292	6.0005	1427.05
	−50	0.25	17.192	5.4428	1590.87
β	50	6.30	17.390	5.3300	810.87
	25	6.15	17.391	5.2140	1120.05
	0	0.6	17.392	6.503	1326.76
	−25	0.45	17.393	6.995	1498.79
	−50	0.3	17.394	6.4999	1589.92

1.6.1 Results analysis

- The overall cost falls if the inflation reduction parameter (r) falls, which is in line with reality.
- If we decrease the value of demand parameters (a) (α) and (β), then we observe that the demand for the products decreases, which implies the increasing value of total cost.
- As values of C_s C_l, and C_h, decrease, we observe that the total cost decreases.

1.6.2 Managerial implications

The managerial implications of these results provide valuable insights into inventory and cost management strategies in an inflation-sensitive environment:

- **Inflation Sensitivity**: As inflation $(\text{parameter}\, r)$ rises, the total cost decreases. This insight suggests that managers can potentially leverage inflationary conditions to reduce overall costs, perhaps by optimizing inventory purchasing and holding strategies to capitalize on the cost-lowering effect of inflation on inventory costs.

- **Demand-Related Sensitivity**: The parameter a directly increases demand, which raises the total cost. Managers should monitor this parameter to ensure that demand growth is balanced with profitability. Higher demand raises costs, but decreased demand can reduce profits, so managing inventory and pricing to optimize demand-related expenses is key.
- **Customer Returns Impact**: An increase in the customer returns parameter (δ) raises the total cost, signaling the need for quality control, customer satisfaction, and return reduction strategies. Reducing returns will help control costs, making the inventory system more efficient and profitable.

1.7 CONCLUSION

The proposed model addresses an inventory management problem under inflation, incorporating scenarios where stock shortages occur and are managed with partial backlogging and a constant rate of deterioration. Demand in this model is influenced by three key independent factors: price, time, and stock level, making it adaptable to diverse, real-world demand situations.

Numerical-1 and the concavity graph shown in Figure 1.2 represent the optimization of total cost, which is matched with real-word scenarios, where the optimal values of the decision variable are $t_1^* = 6.5843$,

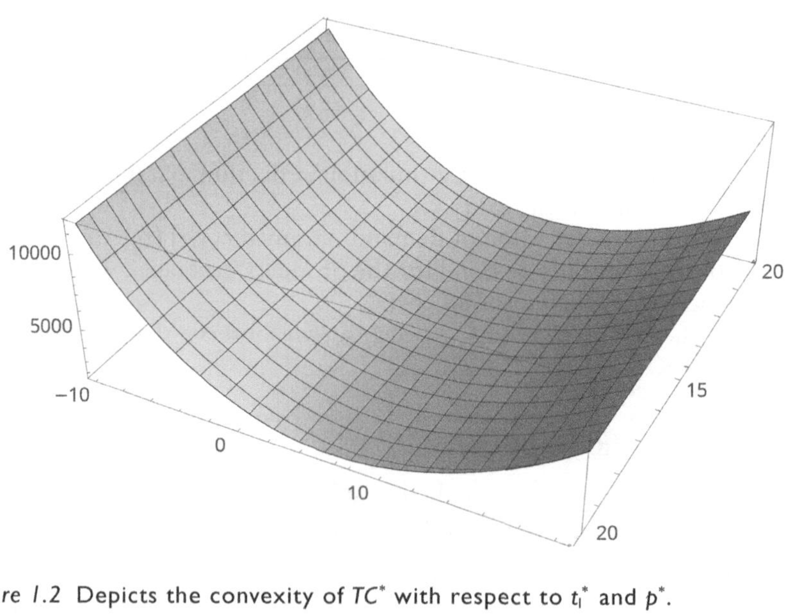

Figure 1.2 Depicts the convexity of TC^* with respect to t_1^* and p^*.

$p^* = 17.392 Q^* = 303.306$, and $TC^* = 1326.76$. This flexibility allows the model to capture various market dynamics effectively. Future extensions could enhance the model by introducing varying backlogging rates and alternative deterioration patterns, such as Weibull distribution or time-dependent rates. Additionally, integrating trade credit policies with permissible payment delays could provide further relevance in real-world financial settings.

REFERENCES

Aarya, D.D. and Kumar, M. (2018) 'A production inventory model with selling price and stock sensitive demand under partial backlogging', *International Journal of Mathematics in Operational Research*, Vol. 12, No. 3, pp. 350–363.

Abad, P.L. (1996) 'Optimal pricing and lot sizing under conditions of perishability and partial backordering', *Management Science*, Vol. 42, No. 8, pp. 1093–1104.

Abad, P.L. (2001) 'Optimal price and order size for a reseller under partial backordering', *Computers & Operations Research*, Vol. 28, No. 1, pp. 53–65.

Buzacott, J.A. (1975) 'Economic order quantities with inflation', *Operational Research Quarterly*, Vol. 26, No. 3, pp. 553–558.

Dye, C.Y. (2007) 'The joint pricing and ordering policy for a deteriorating inventory with price dependent demand and partial backlogging', *Omega*, Vol. 35, No. 2, pp. 184–189.

Goel, R., Singh, A.P. and Sharma, R. (2015) 'Supply chain model with stock dependent demand, quadratic rate of deterioration with allowable shortage', *International Journal of Mathematics in Operational Research*, Vol. 7, No. 2, pp. 156–177.

Goyal, S.K., Hariga, M.A. and Alyan, A. (1996) 'The trended inventory lot sizing problem with shortages and a linear trend in demand', *Journal of the Operational Research Society*, Vol. 47, No. 10, pp. 1286–1295.

Hsieh, T.P., Dye, C.Y. and Ouyang, L.Y. (2010) 'Optimal lot size for an item with partial backlogging rate when demand is stimulated by inventory above a certain stock level', *Mathematical and Computer Modelling*, Vol. 51, No. 1–2, pp. 13–32.

Jaggi, C.K. and Khanna, A. (2010) 'Supply chain model for deteriorating items with stock-dependent consumption rate and shortages under inflation and permissible delay in payment', *International Journal of Mathematics in Operational Research*, Vol. 12, No. 4, pp. 491–514.

Jain, R. and Singh, S.R. (2011) 'Inflationary implications on an inventory with expiration date, capital constraint and uncertain lead time in a multi-echelon supply chain', *International Journal of Procurement Management*, Vol. 4, No. 4, pp. 419–432.

Kumar, N., Singh, S.R. and Kumari, R. (2012) 'Three echelon supply chain inventory model for deteriorating items with limited storage facility and lead–time under inflation', *International Journal of Services and Operations Management*, Vol. 13, No. 1, pp. 98–118.

Lee, Y.P. and Dye, C.Y. (2012) 'An inventory model for deteriorating items under stock-dependent demand and controllable deterioration rate', *Computers and Industrial Engineering*, Vol. 63, No. 2, pp. 474–482.

Ouyang, L.Y., Heish, T.P., Dye, C.Y. and Chang, H.C. (2003) 'An inventory model for deteriorating items with stock dependent demand under the condition of inflation and time value of money', *The Engineering Economist*, Vol. 48, No. 1, pp. 52–68.

Padmanabhan, G. and Virat, P. (1995) 'EOQ models for perishable items under stock dependent selling rate', *European Journal of Operational Research*, Vol. 86, No. 2, pp. 281–292.

Palanivel, M., Sundararajan, R. and Uthayakumar, R. (2016) 'Two-warehouse inventory model with non-instantaneously deteriorating items, stock-dependent demand, shortages and inflation', *Journal of Management Analytics*, Vol. 3, No. 2, pp. 152–173.

Rathore, H., Meedal, T., Bhawaria, S. et al. (2024) 'An optimal inventory policy with ramp type demand rate and time varying holding cost in two-warehousing environment', *Proceedings of the Indian National Science Academy*. https://doi.org/10.1007/s43538-024-00242-9

Rezagholifam, M., Sadjadi, S.J., Heydari, M. and Karimi, M. (2022) 'Optimal pricing and ordering strategy for non-instantaneous deteriorating items with price and stock sensitive demand and capacity constraint', *International Journal of Systems Science: Operations & Logistics*, Vol. 9, No. 1, pp. 121–132. https://doi.org/10.1080/23302674.2020.1833259

Sarkar, B., Sana, S.S. and Chaudhuri, K. (2010) 'A finite replenishment model with increasing demand under inflation', *International Journal of Mathematics in Operational Research*, Vol. 2, No. 3, pp. 347–385.

Sarkar, B., Sana, S.S. and Chaudhuri, K. (2011) 'An imperfect production process for time varying demand with inflation and time value of money – an EMQ model', *Expert Systems with Applications*, Vol. 38, No. 11, pp. 13543–13548.

Singh, S.R., Kumar, N. and Kumari, R., (2010a) 'An inventory model for deteriorating items with shortages and stock-dependent demand under inflation for two-shops under one management', *Opsearch*, Vol. 47, No. 4, pp. 311–329.

Singh, S.R., Kumari, R. and Kumar, N. (2010b) 'Replenishment policy for non-instantaneous deteriorating items with stock-dependent demand and partial back logging with two-storage facility under inflation', *International Journal of Operations Research and Optimization*, Vol. 1, No. 1, pp. 161–179.

Singh, C. and Singh, S.R. (2011) 'Imperfect production process with exponential demand rate, Weibull deterioration under inflation', *International Journal of Operational Research*, Vol. 12, No. 4, pp. 430–445.

Singh, S.R. and Sharma, S. (2014) 'Optimal trade-credit policy for perishable items deeming imperfect production and stock dependent demand', *International Journal of Industrial Engineering Computations*, Vol. 5, No. 1, pp. 151–168.

Thilagavathi, R., Viswanath, J., Cepova, L. and Schindlerova, V. (2022) 'Effect of inflation and permitted three-slot payment on two-warehouse inventory system with stock-dependent demand and partial backlogging', *Mathematics*, Vol. 10, p. 3943. https://doi.org/10.3390/math10213943

Tripathi, R.P. and Mishra, S.M. (2013) 'Optimal ordering policy with stock-dependent demand rate under permissible delay in payments', *Iranian Journal of Operations Research*, Vol. 4, No. 2, pp. 209–217.

Udayakumar, R., Geetha, K.V. and Sana, S.S. (2020) 'Economic ordering policy for non- instantaneous deteriorating items with price and advertisement dependent demand and permissible delay in payment under inflation', *Mathematical Methods in the Applied Sciences*, Vol. 44, No. 9, pp. 7697–7721.

Yang, H.-L. (2006) 'Two-warehouse partial backlogging inventory models for deteriorating items under inflation', *International Journal of Production Economics*, Vol. 103, pp. 362–370. https://doi.org/10.1016/j.ijpe.2005.09.003

Yang, H.L. and Chang, C.T. (2013) 'A two-warehouse partial backlogging inventory model for deteriorating items with permissible delay in payment under inflation', *Applied Mathematical Modelling*, Vol. 37, No. 5, pp. 2717–2726, ISSN 0307-904X. https://doi.org/10.1016/j.apm.2012.05.008

Zhou, Y.W., Lau, H.S. and Yang, S.L. (2004) 'A finite horizon lot-sizing problem with time-Varying deterministic demand and waiting-time-dependent partial backlogging', *International Journal of Production Economics*, Vol. 91, No. 2, pp. 109–119.

Chapter 2

An inventory model with learning effect under reliability-dependent demand

Mahesh Kumar Jayaswal and Faizan Ahmad Khan

2.1 INTRODUCTION

The quality of items damages over time due to the presence of deterioration, which cannot be ignored when the items are deteriorating. Most of the items like electronic items, mobile phones, and chemicals, lose their quality due to the presence of deterioration over time. Depreciation is one of the types of deterioration. The impact of decaying items on fashion was proposed by Whitin (1957). Ghare and Schrader (1963) developed the economic order quantity (EOQ) model where the deterioration rate is constant. Demand rate is the most essential factor in the classical inventory model for the mathematical model. Actually, demand rate changes with time in different situations. A lot of companies face more competition to arrange good quality items for their clients and receive good feedback in the modern market. The good quality means high reliability. Some renowned authors derived a model for inventory where the demand rate is a function of time. Wu (2001) developed a mathematical model for decaying items where the deteriorating rate follows a Weibull distribution and the demand rate follows ramp-type pattern. When a product is reliable, its demand is always present in the market and sometimes this product may not be available when needed, leading to shortages in the shop. In this view, Teng et al. (2002) proposed an inventory model for deteriorating items where the demand rate follows partially backlogging.

Khara et al. (2017) proposed an inventory model for a production system where lots have some defective items. Here, some literature includes models where the demand rate is a function of the selling price and reliability of the items. Mahapatra et al. (2017) derived an EOQ model with deterioration under partial backlogging and reliability where the demand rate is a linear function of time. Shah and Naik (2018) considered an EOQ inventory model under a discount scheme where the demand rate is a function of the selling price. Now moving to the direction of learning concept, it is a mathematical concept, which minimizes the inventory cost. After that, a lot of inventory models with a learning effect with different approaches have been presented. In this flow, Wright (1936) introduced the concept of learning and presented a mathematical model in which the airplane cost is affected

 DOI: 10.1201/9781003521389-2

by learning effects. Jayaswal et al. (2019a) proposed an EOQ model with credit financing under learning effects for defective items.

Jayaswal et al. (2021) presented a model with credit financing and learning effects under an inspection process for decaying defective quality items. Mittal et al. (2021) assumed an EOQ model with shortages and credit financing under learning effects, where demand is a function of time and the deterioration rate follows the Weibull distribution. We are motivated by some literature like Malik and Singh (2013), Malik et al. (2017a, 2017b), Malik and Sharma (2011), Vikram et al. (2016), and Singh and Malik (2010), and these authors gave an inventory model with different approaches for deteriorating items under various conditions. In this direction, Malik et al. (2024) proposed a green supply chain model with inflation and preservation under a trade-credit policy. Yadav et al. (2022) developed a fuzzy-based inventory model for deteriorating items under a variable demand rate. Singh and Malik (2010) assumed an EOQ model with variable demand rate under some realistic situations. Malik and Garg (2021) assumed an inventory model with the impact of fuzzy concept under two warehouses. Alamri et al. (2023) developed a supply chain model with model with the effect of fuzzy system for imperfect quality items under trade credit policy. Alsaedi et al. (2023) proposed a supply chain model with the effect of learning for imperfect items under carbon emissions. In this order, Jayaswal et al. (2019b) presented an ordering policies with the effect of learning for imperfect quality items under trade credit policy. The rest of the paper is presented in the following sections.

2.2 ASSUMPTIONS AND NOTATIONS

2.2.1 Assumptions

1. The continuity of replacement is allowed.
2. Shortages included in this model.
3. The time horizon plane has been considered finite.
4. The learning effect is involved in holding, ordering, shortage, and deterioration costs.
5. Demand rate is a function of reliability and time.
6. Lots have a constant deterioration rate.
7. Lead time is not included in this model.

2.2.2 Notations

a: Parameter for demand rate
$I(t)$: Inventory level in the time interval $[0, t_2]$
I_{max}: Optimal inventory level at $t = t_1$ (unit)
t_1: Time where shortages are completely backlogged (deceision variable) (year)

t_2: Cycle length (deceision variable)(year)
α: learning parameter
n: Number of shipment
$OA = A_0 + \frac{A_1}{n^\alpha}$, where A_0 and A_1 partially fixed ordering cost $(\$ / \text{order})$
$h = h_0 + \frac{h_1}{n^\alpha}$, where h_0 and h_1 partially fixed holding cost $(\$ / \text{unit} / \text{year})$
$d = d_0 + \frac{d_1}{n^\alpha}$, where d_0 and d_1 partially fixed deteriorating cost $(\$ / \text{unit})$
$s = s_0 + \frac{s_1}{n^\alpha}$, where s_0 and s_1 partially fixed shortage cost $(\$ / \text{unit} / \text{year})$
θ: Rate of deterioration
M: Marketing expenditure
w: Reliability of items of this inventory level, that is, w% of good quality items present
$D(t) = atw^t$: Rate of demand

2.3 MATHEMATICAL FORMULATION

Initially, the inventory stock will be zero when $t = 0$. On account of advance booking of items, shortages occur in the time interval $[0, t_1]$. When $t = t_1$, shortages are completely backlogged. After that the maximum inventory level reduces due to demand and deterioration in the interval $[t_1, t_2]$. The model figure has been shown in Figure 2.1.

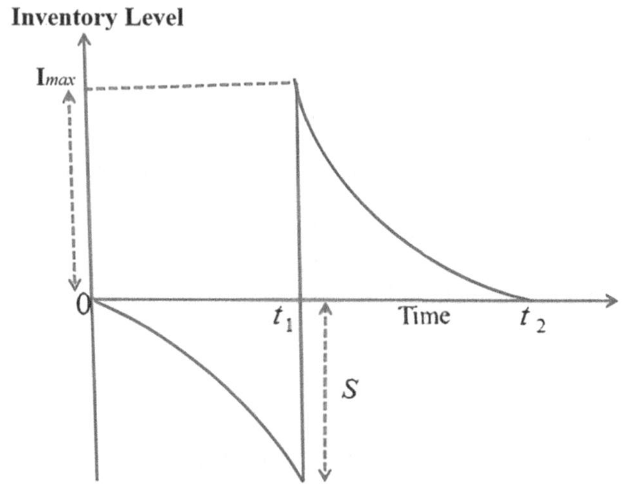

Figure 2.1 Inventory with time.

Suppose that $I(t)$ be the inventory level at time t and varies with respect to time, then we can write in the form of an ordinary differential equation with a boundary condition, which is given in equation (2.1)

$$\frac{dI(t)}{dt} = -D(t), 0 \le t \le t_1 \tag{2.1}$$

and

$$\frac{dI(t)}{dt} + \theta I(t) = -D(t),\ t_1 \le t \le t_2$$

with boundary condition,

$$I(0) = 0, I(t_2) = 0. \tag{2.2}$$

Now, we have the solution to the equations (2.1) and (2.2), which are given as follows:

$$I(t) = \frac{-aw^t\left[\log[w]t - 1\right] - a}{\log[w]^2}, 0 \le t \le t_1 \tag{2.3}$$

and

$$I(t) = \frac{aw^{t_1} + aw^t - atw^t\left(\log[w] + \theta\right) - at_1w^{t_1}\left(\log[w] + \theta\right)}{\log[w]^2},\ t_1 \le t \le t_2 \tag{2.4}$$

From equation (2.4), the inventory will be maximum when $t = t_1$ and its maximum value is

$$I_{\max} = \frac{aw^{t_1} + aw^{t_2} - \left(at_1w^{t_1} + at_2w^{t_2}\right)\left(-\log[w] + \theta\right)}{\left(\theta + \log[w]\right)^2} \tag{2.5}$$

The demand for backlogged is as follows:

$$S = \int_0^{t_1} D(t)dt = \int_0^{t_1} atw^t dt = \frac{aw^t\left[\log[w]t_1 - 1\right]}{\log[w]^2} + \frac{a}{\log[w]^2} \tag{2.6}$$

The inventory level at initial $I_0 = I_{\max} + S$ can be written from equations (2.5) and (2.6):

$$I_0 = \frac{aw^{t_1} + aw^{t_2} - \left(at_1w^{t_1} + at_2w^{t_2}\right)\left(-\log[w] + \theta\right)}{\left(\theta + \log[w]\right)^2} + \frac{aw^t\left[\log[w]t_1 - 1\right]}{\log[w]^2} + \frac{a}{\log[w]^2} \tag{2.7}$$

Now, the total cost of this inventory model is the sum of ordering cost, holding cost, and the cost of deterioration, which are given as follows:

$$\text{Ordering cost,} \quad OA = A_o + \frac{A_1}{n^{\alpha}} \tag{2.8}$$

Holding cost,

$$IHC = h_o + \frac{h_1}{n^{\alpha}}\left[\int_{t_1}^{t_2} I(t)\,dt\right] = \left(h_o + \frac{h_1}{n^{\alpha}}\right)$$

$$\left[\int_{t_1}^{t_2} \frac{aw^{t_1} + aw^{t_2} - \left(at_1w^{t_1} + at_2w^{t_2}\right)\left(-\log[w] + \theta\right)}{\left(\theta + \log[w]\right)^2}\,dt\right]$$

$$= \frac{\left(h_o + \frac{h_1}{n^{\alpha}}\right)a}{\left(\theta^2 + \log[w]^2 + 2\theta\log[w]\right)} \tag{2.9}$$

$$\left[\begin{array}{l}\dfrac{t_2w^{t_2}\log[w] + w^{t_2} - w^{t_1} - w^{t_1}t_1}{\log[w]} + \\ \left(\log[w] + \theta\right)\left(-w^{t_1} + \log[w]t_1w^{t_1}\right) + \\ \dfrac{w^2 - w^2t_2\log[w]\left(1 + t_2\log[w]\right) + t_1t_2\log[w]^2}{\log[w]^2}\end{array}\right]$$

Deterioration cost

$$DC = \left(d_0 + \frac{d_1}{n^{\alpha}}\right)\left[\begin{array}{l}\dfrac{aw^{t_1} + aw^{t_2} - \left(at_1w^{t_1} + at_2w^{t_2}\right)\left(-\log[w] + \theta\right)}{\left(\theta + \log[w]\right)^2} \\ -\displaystyle\int_{t_1}^{t_2} D(t)\,dt\end{array}\right]$$

$$= \left(d_0 + \frac{d_1}{n^{\alpha}}\right)\left[\begin{array}{l}\dfrac{aw^{t_1} + aw^{t_2} - \left(at_1w^{t_1} + at_2w^{t_2}\right)\left(-\log[w] + \theta\right)}{\left(\theta + \log[w]\right)^2} \\ -\dfrac{aw^{t_1} + aw^{t_2} - \log[w]\left(at_1w^{t_1} + at_2w^{t_2}\right)}{\log[w]^2}\end{array}\right] \tag{2.10}$$

Shortage cost

$$SC = \left(s_0 + \frac{s_1}{n^{\alpha}}\right)\left[\int_0^{t_1} D(t)\,dt\right]$$

$$= \left(\left(s_0 + \frac{s_1}{n^{\alpha}}\right)\right)\left[\frac{aw^t\left[\log[w]t_1 - 1\right]}{\log[w]^2} + \frac{a}{\log[w]^2}\right] \tag{2.11}$$

Marketing cost,

$$MQ = m\left(\frac{aw^{t_1} + aw^{t_2} - \left(at_1w^{t_1} + at_2w^{t_2}\right)\left(-\log\left[w\right]+\theta\right)}{\left(\theta + \log\left[w\right]\right)^2} + \frac{aw^t\left[\log\left[w\right]t_1 - 1\right]}{\log\left[w\right]^2} + \frac{a}{\log\left[w\right]^2}\right) \tag{2.12}$$

Now, the total average cost is the sum of ordering cost, holding cost, the cost of deterioration, as well as marketing cost, which are given as follows:

$$\Psi\left(t_1,t_2\right) = \frac{OA + IHC + DC + SC + MQ}{t_2} \tag{2.13}$$

The values of ordering cost, holding cost, deterioration cost, shortage cost, as well as marketing cost from equations (2.8), (2.9), (2.10), (2.11), and (2.12), are substituted in equation (2.13), we get

$$\Psi\left(t_1,t_2\right) = \frac{\begin{gathered}\left(A_o + \frac{A_1}{n^\alpha}\right) + \left(\frac{\left(h_o + \frac{h_1}{n^\alpha}\right)\alpha}{\left(\theta^2 + \log[w]^2 + 2\theta\log[w]\right)}\left[\frac{t_2w^{t_2}\log[w] + w^{t_2} - w^{t_1} - w^{t_1}t_1}{\log[w]} + \frac{\left(\log[w]+\theta\right)\left(-w^{t_1} + \log[w]t_1w^{t_1}\right) + w^2 - w^2t_2\log[w]\left(1 + t_2\log[w]\right) + t_1t_2\log[w]^2}{\log[w]^2}\right]\right) + \\ \left(d_0 + \frac{d_1}{n^\alpha}\right)\left[\frac{\alpha w^{t_1} + \alpha w^{t_2} - \left(\alpha t_1w^{t_1} + \alpha t_2w^{t_2}\right)\left(-\log[w]+\theta\right)}{\left(\theta + \log[w]\right)^2} - \frac{\alpha w^{t_1} + \alpha w^{t_2} - \log[w]\left(\alpha t_1w^{t_1} + \alpha t_2w^{t_2}\right)}{\log[w]^2}\right] + \\ \left(\left(s_0 + \frac{s_1}{n^\alpha}\right)\right)\left[\frac{\alpha w^t\left[\log[w]t_1 - 1\right]}{\log[w]^2} + \frac{\alpha}{\log[w]^2}\right] \\ + m\left(\frac{\alpha w^{t_1} + \alpha w^{t_2} - \left(\alpha t_1w^{t_1} + \alpha t_2w^{t_2}\right)\left(-\log[w]+\theta\right)}{\left(\theta + \log[w]\right)^2} + \frac{\alpha w^t\left[\log[w]t_1 - 1\right]}{\log[w]^2} + \frac{\alpha}{\log[w]^2}\right)\end{gathered}}{t_2} \tag{2.14}$$

2.4 SOLUTION METHOD

For optimal values of t_1 and t_2, if we take $\frac{\partial\Psi(t_1,t_2)}{\partial t_1}=0$, and $\frac{\partial\Psi(t_1,t_2)}{\partial t_2}=0$, for the convexity of the total average cost, the following condition must be satisfied:

$$\left(\frac{\partial^2\Psi(t_1,t_2)}{\partial t_1\partial t_2}\right)^2-\left(\frac{\partial^2\Psi(t_1,t_2)}{\partial t_1^2}\right)\left(\frac{\partial^2\Psi(t_1,t_2)}{\partial t_2^2}\right)\geq 0$$

$$and\quad\left(\frac{\partial^2\Psi(t_1,t_2)}{\partial t_1^2}\right)\geq 0,\quad\left(\frac{\partial^2\Psi(t_1,t_2)}{\partial t_2^2}\right)\geq 0,$$

Note: The average total cost is calculated with the help of Mathematica Software for the sensitivity analysis.

2.5 NUMERICAL EXAMPLE

$a=25, w=80\%, \alpha=0.08, h_0=\$2/\text{unit}/\text{year},\ h_1=\$4/\text{unit}/\text{year},$
$A_0=\$3/\text{unit}/\text{year},$
$A_1=\$12/\text{unit}/\text{year},\ d_0=\$5/\text{unit}/\text{year},\ d_1=\$10/\text{unit}/\text{year},$
$s_0=\$5/\text{unit},\ s_1=\$2/\text{unit}, M=\$0.70,$
$\theta=0.01,\ ^*=5,$ then we got optimal procurement time $t_1^*=0.2439$ year, cycle time $t_2^*=0.2372$ year and optimal total average cost $\Psi(t_1^*,t_2^*)=1318\$.$

2.6 SENSITIVITY ANALYSIS

Sensible inspection is accomplished to know the hardiness of the model on the affected parameters:

2.6.1 Impact of shipment

Table 2.1 shows that the aggregate of shipment increases from 1 to 5, and then the shortage period time and total average cost decrease while the cycle length increases. It advises that the buyer can manage his total average cost with respect to shipments. The intensity level of total average cost is not attained till fifth shipment, whereas it can be reached in the fifth shipment if the learning rate is 0.08.

2.6.2 Impact of learning

Table 2.2 and Figure 2.2 state that if the learning rate rises from 0.04 to 0.08, then the cycle length, shortage period time, and total average cost

Table 2.1 Effect of shipment on procurement time, cycle time, and total average cost

Number of shipment n	*Procurement time t_1 (year)*	*Cycle time t_2 (year)*	*Total average cost $\Psi(t_1, t_2)$ (\$)*
1	0.2465	0.2345	1352
2	0.2460	0.2350	1342
3	0.2452	0.2356	1334
4	0.2448	0.2365	1326
5	0.2439	0.2372	1318

Table 2.2 Effect of shipment on procurement time, cycle time and total average cost

Learning factor α	*Procurement time t_1 (year)*	*Cycle time t_2 (year)*	*Total average cost $\Psi(t_1, t_2)$ (\$)*
0.04	0.2765	0.2655	1462
0.05	0.2660	0.2550	1442
0.06	0.2552	0.2456	1385
0.07	0.2448	0.2387	1326
0.08	0.2439	0.2372	1318

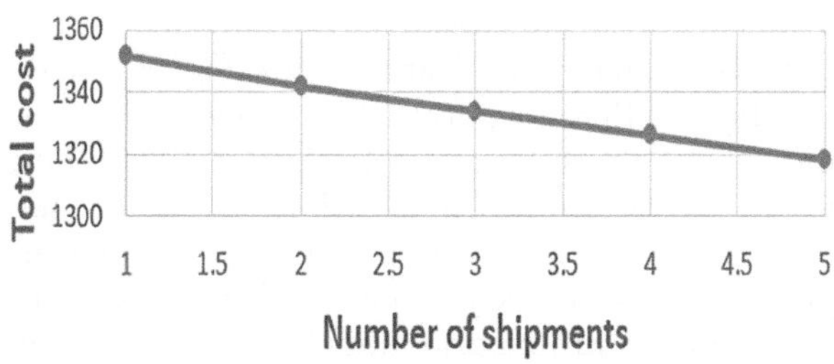

Figure 2.2 Impact of shipment on the total inventory cost.

decrease. It gives more information to the decision-maker, and it helps them to earn more profit for the company.

2.6.3 Impact of parameter of demand rate

Table 2.3 shows that when the parameter of demand rate decreases, then the shortage period time, cycle length, and total average cost decrease.

2.7 CONCLUSION

This chapter explains the ordering policy with the effect of learning for decaying items under shortages, where the demand rate is a function of the item reliability and time factor. The novelty of this chapter is that the total

Table 2.3 Effect of demand parameter on procurement time, cycle time, and total average cost

Demand parameter a	*Procurement time* t_1 (year)	*Cycle time* t_2 (year)	*Total average cost* $\Psi(t_1, t_2)$ *($)*
200	0.2975	0.2885	2636
150	0.2860	0.2650	2595
125	0.2752	0.2648	2485
100	0.2548	0.2487	2326
50	0.2439	0.2372	1318

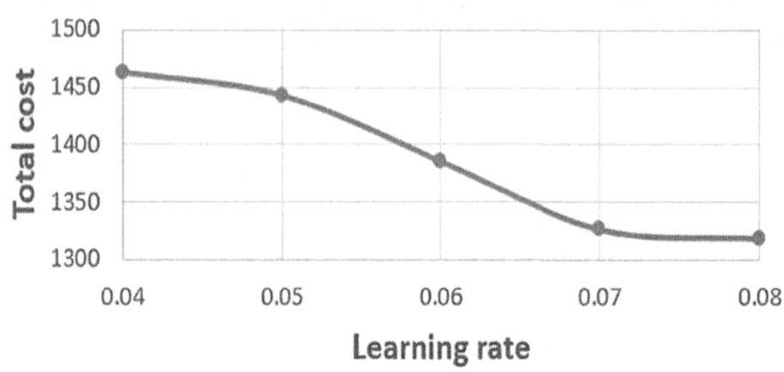

Figure 2.3 Impact of learning rate on the total inventory cost.

average cost can be managed by learning effects. Moreover, as shown in Table 2.3, it can be said that the parameters of demand rate, effect of learning, and shipment have a positive effect on the ordering policy. The ordering cost, holding cost, deterioration cost, and shortages cost reduce as shipment increases due to present of leaning effect and also shown in Figure 2.3. It is suggested that the present study is only valid for buyer's ordering policy and it is not valid for the production model. This chapter enhances a more sensible position, for instance in supply chain management and two-level trade-credit policies.

REFERENCES

Alamri, O. A., Jayaswal, M. K., & Mittal, M. (2023), "A supply chain model with learning effect and credit financing policy for imperfect quality items under fuzzy environment," *Axioms*, 12(3), 260.

Alsaedi, B. S., Alamri, O. A., Jayaswal, M. K., & Mittal, M. (2023), "A sustainable green supply chain model with carbon emissions for defective items under learning in a fuzzy environment," *Mathematics*, 11(2), 301.

Ghare, P. M., & Schrader, G. P. (1963), "A model for exponential decaying inventory," *Journal of Industrial Engineering*, 14(3), 238–243.

Jayaswal, M., Sangal, I. S. H. A., Mittal, M., & Malik, S. (2019a), "Effects of learning on retailer ordering policy for imperfect quality items with trade credit financing," *Uncertain Supply Chain Management*, 7(1), 49–62.

Jayaswal, M. K., Mittal, M., & Sangal, I. (2021), "Ordering policies for deteriorating imperfect quality items with trade-credit financing under learning effect," *International Journal of System Assurance Engineering and Management*, 12(1), 112–125.

Jayaswal, M. K., Sangal, I., & Mittal, M. (2019b, February), "Learning effect on stock-policies with Imperfect quality and deteriorating Items under trade credit," In *2019 Amity International Conference on Artificial Intelligence (AICAI)* (pp. 499–504). IEEE, Dubai.

Khara, B., Dey, J. K., & Mondal, S. K. (2017), "An inventory model under development cost-dependent imperfect production and reliability-dependent demand," *Journal of Management Analytics*, 4(3), 258–275.

Mahapatra, G. S., Adak, S., Mandal, T. K., & Pal, S. (2017), "Inventory model for deteriorating items with time and reliability dependent demand and partial backorder," *International Journal of Operational Research*, 29(3), 344–359.

Malik, A. K., Chakraborty, D., Bansal, K. K., & Kumar, S. (2017a), "Inventory model with quadratic demand under the two warehouse management system," *International Journal of Engineering and Technology*, 9(3), 2299–2303.

Malik, A. K., & Garg, H. (2021), "An improved fuzzy inventory model under two warehouses," *Journal of Artificial Intelligence and Systems*, 3, 115–129.

Malik, A. K., Kumar, S., & Kumar, S. (2017b), "Two warehouses inventory model with stock-dependent demand and maximum life time," *International Journal of Physical Sciences*, 29, 309–318.

Malik, A. K., Pal, S., Jain, D. R., & Modibbo, U. M. (2024), "A green supply chain model for decaying items with learning and inflation under the strategy of trade credit and preservation technology," *Process Integration and Optimization for Sustainability*, 8(5), 1483–1492.

Malik, A. K., & Sharma, A. (2011), "A two warehouses inventory model for linear deterioration with exponentially demand rate," *Journal of Ultra Scientist of Physical Sciences*, 23(2), 577–784.

Malik, A. K., & Singh, Y. (2013), "A fuzzy mixture two warehouse inventory model with linear demand," *International Journal of Application or Innovation in Engineering and Management*, 2(2), 180–186.

Mittal, M., Jayaswal, M. K., & Kumar, V. (2021), "Effect of learning on the optimal ordering policy of inventory model for deteriorating items with shortages and trade-credit financing," *International Journal of System Assurance Engineering and Management*, 13(2), 913–924.

Shah, N. H., & Naik, M. K. (2018), "Inventory policies for price-sensitive stock-dependent demand and quantity discounts," *International Journal of Mathematical, Engineering and Management Sciences*, 3(3), 245–257.

Singh S. R., & Malik A. K. (2010), "Inventory system for decaying items with variable holding cost and two shops," *International Journal of Mathematical Sciences*, 9(3–4), 489–511.

Teng, J. T., Chang, H. J., Dye, C. Y., & Hung, C. H. (2002), "An optimal replenishment policy for deteriorating items with time-varying demand and partial backlogging," *Operations Research Letters*, 30(6), 387–393.

Vikram, V., Ajay, T., Chandra, S., & Malik, A. K. (2016), "A trade credit inventory model with multivariate demand for non-instantaneous decaying products," *Indian Journal of Science and Technology*, 9(15), 1–6.

Whitin, T. M. (1957), *Theory of inventory management*, Princeton University Press, Princeton, NJ, 62–72.

Wright, T. P. (1936), "Factors affecting the cost of airplanes," *Journal of the Aeronautical Sciences*, 3(4), 122–128.

Wu, K. S. (2001), "An EOQ inventory model for items with Weibull distribution deterioration, ramp type demand rate and partial backlogging," *Production Planning and Control*, 12(8), 787–793.

Yadav, V., Chaturvedi, B. K., & Malik, A. K. (2022), "Development of fuzzy inventory mode under decreasing demand and increasing deterioration rate," *International Journal on Future Revolution in Computer Science & Communication Engineering*, 8, 1–7.

Chapter 3

Inventory model for inevitable stock with deteriorating objects requirement dependent on time precipitously and stationary holding with cost debilitation rate

Animesh Kumar Sharma

3.1 INTRODUCTION

Every specialist aims to reduce the total cost of stock by establishing an inventory policy that ideally locks in inventory. Inventory professionals examine trade credit policies in order to ensure they are strong enough to achieve this objective. This leads to a decrease in the retailer's costs associated with buying and managing suppliers. Normally, the trader agrees to the proposal as he needs more money to settle with the vendor and quickly close the account. The profit account of the seller can be increased by adding the revenue earned from product sales during this period. Interest can be earned by merchants on the money they deposit. The seller returns the outstanding share once the debt has been settled. Hence, the supplier's trade credit term is a vital marketing tool to attract the retailer. The seller has the option to use it as a different reward system. Inventory is a product that needs to be organized in a way that maximizes profits and minimizes the risk of losses.

Different types of items in inventory fall into two categories: perishable items like bread or fruit that have a limited shelf life and non-perishable items like paper that do not deteriorate over time. Many experts in the field have observed that the demand-based fee remains constant, with some considering the price demand as a key element of the era, either as a first- or second-degree exponent characteristic. Jalan et al. (1996) validated their enhanced economic order quantity model with the linear demand or need and Weibull-distributed debilitation. In 1996, exponential debilitation was discussed with primary-level requests. M. Rastogi and P. Kushwaha developed an inventory model utilizing a marketplace that is sensitive to pricing. In a specific stochastic market environment, Aggun integrated weakening and stock levels. Price fluctuations and costs must be taken into account as factors influencing the rate of debilitation.

The linear demand constant deterioration model was created by K. Hou in 2006 and is applied in inventory management. Cheng and G. Wang (2009) created a model for products that have been used up, building on previous work, which includes a trapezoidal demand rate and a constant depletion rate. In 2013, V. K. Mishra developed an exhausted supply using

DOI: 10.1201/9781003521389-3

stationary depletion and demand determined by the first level. B. Mandal and S. Phaujdar (2013) developed an inventory model for declining products or items, stressing the impact of inventory on demand. C. K. Tripathy and U. Mishra created order-level inventories, a time-sensitive Weibull decay method, and demand rate ramping systems in 2015, recognizing the importance of time in production and demand.

Previous models have not yet achieved the integration of supply and selling price with pricing policy in a stochastic market setting. The majority of models did not include the sale price. The selection criteria were determined based on the overall decrease in expenses. In 2016, N. Shah, U. Chaudhari, and Mudul Y. Jani collaborated on addressing time- and price-sensitive demand to develop an integrated manufacturing inventory. They further developed methods to preserve perishable goods over time. In 2019, N. Shah and M. Naik developed an EOQU model for damaged items. (EOQU-Economic Order Quantity with Upfront payment) .If the demand is responsive to price changes, it is recommended to pay the full amount in advance in order to benefit from discounts. A. H. M. Mashud (2020) created an improved EOQU-deteriorated inventory model despite facing different market types and overall backward congestion.

Therefore, we considered the rate of demand to be dependent on both time and sales price. In a subsequent study, R. Amutha and D. R. Chandrasekaran (2012) created a model for depleted items with partial backlogs, Weibull-distributed degradation, and fluctuating demand. The literature clearly shows that the majority of stock researchers believe that the demand function is a constant and can be differentiated with respect to time. Based on the market analysis, the need for various products, including newly launched items, stays consistent over the entire planning timeframe. It has also been mentioned that the arrangement and inventory shown are usually in proportion. However, it is not believable to think that demand cannot slowly increase or decrease. Once the customers in the market approve the product, the demand for it becomes steady. Demand will remain stable for the rest of the planning period.

In 2017, M. Maragatham and R. Palani developed an inventory system that was deterministic and involved factors such as deterioration and demand rate, time-proportional deterioration rate, as well as costs for inventory holding, ordering, and deterioration. The policy rule that minimizes total cost was discovered. This model of inventory takes into account perishable items in a single storage facility. This model resulted in delays in lead times and a full backlog. The mathematical inventory model includes holding costs that depend on stock levels, demand rates that vary with stock levels, and flexible terminal constraints. This permits the reordering to take place in either of two situations: (1) The ordering points could be less than or equal to zero because of N_0. If there is insufficient supply and a backlog, the reorder point is below zero. (2) The formula N_0 was developed by Shuhua Zhang and Longzhou Cao. This implies that there may be a shortage in 2022 or the reorder level could be set at zero. They recognized the uniqueness and importance of the reorder level, optimal order-up-to level, and investment in preservation technology.

Therefore, in this chapter, a deterministic type inventory model is established for item deterioration with a constant debilitation rate, specifically for unit objects with a time-variable demand rate, where the rate of "*m*" is raised to the "*n*" th power. Shortfalls are allowed and fully backlogged, and costs based on such holdings are considered constant.

3.2 DEVELOPMENT AND EXPLICATION OF THE MATHEMATICAL MODEL

3.2.1 Notations

- D_r (or D)—Requirement rate
- C_d—Debilitation-based cost
- Oc—Ordering outlay
- C_h (or C_H)—Stock/inventory cost based on holding.
- Qu—Order-based quantity
- θ—Rate of debilitation
- C_P—Purchase cost/unit
- L_Q—Order quantity (for convenience, shortly represented as L)
- Ω(omega) denotes " time" at which the extent of stock or inventory $\to 0$
- T_L—The length of a cycle-time (for convenience, shortly represented as T only)
- Co_S—Shortage cost per unit time (for convenience, shortly represented as C_S)
- TSC—Total inventory cost
- Inv(t)—Inventory level (for convenience, shortly represented as $I(t)$)

3.2.2 Assumptions

The inventory or stock model uses the subsequent assumptions:

- The requirement or demand charge is rationally dependent on time (i.e., nature as m by n)

$$D(t) = \alpha + \beta t^{\frac{m}{n}}; \alpha \geq 0, \beta \geq 0.$$

- Cost based on holding per unit time is constant $\mathrm{Hol}(t) = C_H$, where $Co_H \geq 0$.
- Q or Q_o—Quantity of order per cycle.
- Shortages were permitted and completely backlogged.

- The lead time is null/void, that is, 0 (zero).
- The debilitation rate is $\theta(t) = \gamma, \gamma > 0 \text{ and } \gamma < 1$.
- Oc—Ordering outlay for every item.
- Co_s—Shortages cost per unit time.
- Inventory model deals with only one, that is, single object or article.
- τ—Time exactly when level of stock or inventory reaches zero; also $\tau \geq 0$.
- TSC—Total inventory cost per unit time.
- $\text{Inv}(t)$—The measure level of inventory.
- C_P—Purchasing cost per unit item.

3.2.3 Mathematical formulation

Figure 3.1 represents the extent of inventory or stock at any point in time. This also represents the optimum order quantity for the given time.

The inventory level exhibits a non-negative value and declines in the range $(0, \tau)$, but when the well-known time t equals τ, the above inventory level approaches zero. A shortage begins, making inventory slightly negative in the range (τ, T). The following describes the differential equation for the instantaneous stock or inventory level at a particular moment t:

$$\frac{d}{d\tau} I_1(t) + \mu.(I(t)) = -\left(\alpha + \beta t^{\frac{m}{n}}\right); t \geq 0 \text{ and } t \leq \tau \tag{3.1}$$

$$\frac{d}{d\tau} I_2(t) = -\left(\alpha + \beta t^{\frac{m}{n}}\right); t \geq \tau \text{ and } t \leq T \tag{3.2}$$

where m is an element of a set of non-negative integers with the boundary conditions

$$\text{Inv}_1(t) = \text{Inv}_2(t) = 0 \, \text{at} \, t = \Omega$$

$$I_2(t) = J \, \text{at} \, t = T \, \text{and} \, I_1(t) = E \, \text{at} \, t = 0$$

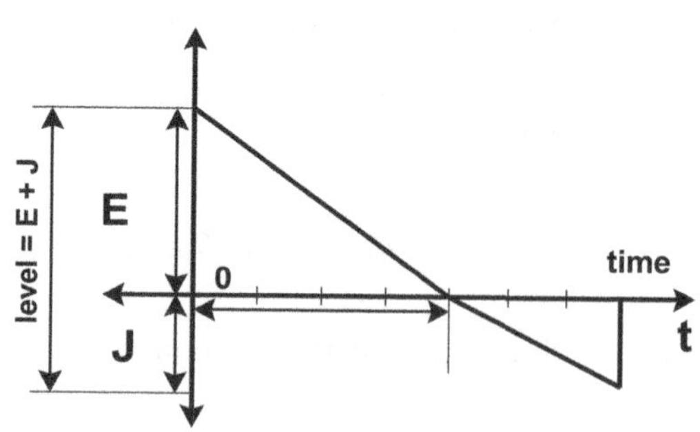

Figure 3.1 Inventory level with time.

On simplifying expression (3.1) and excluding higher powers of t, we get

$$\mathrm{Inv}_1(t)=\begin{bmatrix}\alpha(\tau-t)+\dfrac{\beta n}{m+n}\left(\tau^{\frac{m+n}{n}}-t^{\frac{m+n}{n}}\right)\\ +\dfrac{\mu\alpha}{2}\left(\tau^2-t^2\right)+\dfrac{\beta n\mu}{m+2n}\left(\tau^{\frac{m+n+n}{n}}-t^{\frac{m+n+n}{n}}\right)\\ +\dfrac{\alpha\mu^2}{6}\left(\tau^3-t^3\right)+\dfrac{\beta n\mu^2}{2(m+3n)}\left(\tau^{\frac{m+3n}{n}}-t^{\frac{m+3n}{n}}\right)\\ -\alpha\mu\left(\tau t-t^2\right)-\dfrac{\beta n\mu}{m+n}\left(\tau^{\frac{m+n}{n}}t-t^{\frac{m+2n}{n}}\right)\\ -\dfrac{\alpha\mu^2}{2}\left(\tau^2t-t^3\right)-\dfrac{\beta n\mu^2}{m+2n}\left(\tau^{\frac{m+2n}{n}}t-t^{\frac{m+3n}{n}}\right)\end{bmatrix}+\begin{bmatrix}-\dfrac{\mu^3\alpha}{6}\left(\tau^3t-t^4\right)-\dfrac{\beta n\mu^3}{2(m+3n)}\left(\tau^{\frac{m+3n}{n}}t-t^{\frac{m+4n}{n}}\right)\\ +\dfrac{\mu^2\alpha}{2}\left(\tau t^2-t^3\right)+\dfrac{\beta n\mu^2}{2(m+n)}\left(\tau^{\frac{m+n}{n}}t^2-t^{\frac{m+3n}{n}}\right)\\ +\dfrac{\mu^3\alpha}{4}\left(\tau^2t^2-t^4\right)+\dfrac{\beta n\mu^3}{2(m+2n)}\left(\tau^{\frac{m+2n}{n}}t^2-t^{\frac{m+4n}{n}}\right)\\ +\dfrac{\mu^4\alpha}{12}\left(\tau^3t^2-t^5\right)+\dfrac{\beta n\mu^4}{4(m+3n)}\left(\tau^{\frac{m+3n}{n}}t^2-t^{\frac{m+5n}{n}}\right)\end{bmatrix} \tag{3.3}$$

On simplifying the expression, we obtained

$$I_2(t)=\alpha(\tau-t)+\frac{n\beta}{m+n}\left(\tau^{\frac{m+n}{n}}-t^{\frac{m+n}{n}}\right) \tag{3.4}$$

Now using a particular condition $I_1(t)=E$ at $t=0$, we obtained

$$E=\begin{bmatrix}\alpha\tau+\dfrac{n\beta}{m+n}\tau^{\frac{m+n}{n}}+\dfrac{\alpha\mu}{2}\tau^2+\dfrac{n\beta\mu}{m+2n}\tau^{\frac{m+2n}{n}}\\ +\dfrac{\alpha\mu^2}{6}\tau^3+\dfrac{n\beta\mu^2}{2(m+3n)}\tau^{\frac{m+3n}{n}}\end{bmatrix}$$

Using the initial condition $I_2(t) = J$ at $t = T$ and $\tau = \Omega$, we have

$$J = -\alpha(\Omega - T) - \frac{n\beta}{m+n}\left(\Omega^{\frac{m+n}{n}} - T^{\frac{m+n}{n}}\right)$$

The order-relevant quantity as per the cycle is defined and denoted as follows: $Q = E + J$

$$Q = \begin{bmatrix} \alpha\Omega + \dfrac{n\beta}{m+n}\Omega^{\frac{m+n}{n}} + \dfrac{\alpha\mu}{2}\Omega^2 + \dfrac{n\beta\mu}{m+2n}\Omega^{\frac{m+2n}{n}} + \dfrac{\alpha\mu^2}{6}\Omega^3 \\ + \dfrac{n\beta\mu^2}{2(m+3n)}\Omega^{\frac{m+3n}{n}} - \alpha(\Omega - T) - \dfrac{n\beta}{m+n}\left(\Omega^{\frac{m+n}{n}} - T^{\frac{m+n}{n}}\right) \end{bmatrix} \tag{3.5}$$

Now the inventory cost based on holding per cycle is as follows:

$$IHC = C_H \int_0^{\Omega} I_1(t)\,dt$$

$$IhC = C_H \begin{bmatrix} \dfrac{\alpha}{2}\Omega^2 + \dfrac{\dfrac{1}{\left(\dfrac{1}{n^2}\right)\left(\dfrac{1}{\beta}\right)}}{(m+n)(m+2n)}\Omega^{\frac{m+2n}{n}} + \dfrac{\alpha\mu}{6}\Omega^3 + \dfrac{2n^2\beta\mu}{(m+2n)(m+3n)}\Omega^{\frac{m+3n}{n}} \\ + \dfrac{5\alpha\mu^2}{12}\Omega^4 + \dfrac{3n^2\beta\mu^2}{2(m+3n)(m+4n)}\Omega^{\frac{m+4n}{n}} - \dfrac{n^2\beta\mu}{(m+n)(m+3n)}\Omega^{\frac{m+3n}{n}} \\ - \dfrac{1}{(m+2n)(m+4n)}\alpha\mu^3\Omega^5 - \dfrac{3n^2\beta\mu^3}{2(m+3n)(m+5n)}\Omega^{\frac{m+4n}{n}} \\ + \dfrac{n^2\beta\mu^2}{2(m+n)(m+4n)}\Omega^{\frac{m+4n}{n}} \\ + \dfrac{n^2\beta\mu^3}{(m+2n)(m+5n)}\Omega^{\frac{m+5n}{n}} + \dfrac{\alpha\mu^4}{72}\Omega^6 + \dfrac{3n^2\beta\mu^4}{4(m+3n)(m+6n)}\Omega^{\frac{m+6n}{n}} \end{bmatrix} \tag{3.6}$$

Thus, the overall amount of inventory shortage over the specified period (Ω, T) is defined by the following equation:

$$ISc = -C_S \int_{\Omega}^{T} I_2(t)\,dt$$

$$ISc = -C_S\int_{\Omega}^{T}\left[\alpha(\Omega - t) + \frac{n\beta}{m+m}\left(\Omega^{\frac{m+n}{n}} - t^{\frac{m+n}{n}}\right)\right]dt$$

$$ISc = -C_B\left[\begin{array}{l}\alpha\left(\Omega T - \frac{T^2}{2} - \Omega^2 + \frac{\Omega^2}{2}\right) + \frac{n\beta}{m+n}\\ \left(\Omega^{\frac{n+n}{n}}T - \frac{n}{m+2n}T^{\frac{n+2\pi}{2}} - \Omega^{\frac{n+2\pi}{n}} + \frac{m}{2n}\Omega^{\frac{n+2n}{n}}\right)\end{array}\right]$$

$$ISc = -C_s\left[\begin{array}{l}\alpha\left(\Omega T - \frac{T^2}{2} - \frac{\Omega^2}{2}\right)\\ + \frac{n\beta}{m+n}\left(\Omega^{\frac{n+n}{n}}T - \frac{n}{m+2n}T^{\frac{m+2\pi}{n}} - \frac{(m+n)}{(m+2n)}\Omega^{\frac{n+2n}{n}}\right)\end{array}\right] \tag{3.7}$$

So, inventory debilitation or deterioration cost of the unit item is denoted by the following equation:

$$IdC = C_D\left[E - \int_0^{\Omega}\left(\alpha + \beta t^{\frac{m}{n}}\right)dt\right]$$

$$IdC = C_D\left[\frac{\alpha\gamma\Omega^2}{2} + \frac{n\beta\mu\Omega^{\frac{n+2n}{n}}}{m+2n} + \frac{\alpha\mu^2\Omega^3}{2\times 3} + \frac{n\beta\mu^2\Omega^{\frac{\frac{n+1n}{T}}{5}}}{2(m+3n)}\right] \tag{3.8}$$

Thus, the inventory ordering outlay per order during the open interval $(0, \Omega)$ is as follows:

$$IOC = A \tag{3.9}$$

So, the whole cost/unit time for the cycle is as follows:

$$TSC = \frac{1}{T}\left(\begin{aligned}&\frac{\alpha}{2}\Omega^2 + \frac{n^2\beta}{(m+n)(m+2n)}\Omega^{\frac{n+2n}{n}} + \frac{\alpha\mu}{6}\Omega^3\\ &+\frac{2n^2\beta\mu}{(m+2n)(m+3n)}\Omega^{\frac{m+3n}{n}} + \frac{5\alpha\mu^2}{12}\Omega^4\\ &+\frac{3n^2\beta\mu^2}{2(m+3n)(m+4n)}\Omega^{\frac{n+4\pi}{n}} - \frac{n^2\beta\mu}{(m+n)(m+3n)}\Omega^{\frac{n+3n}{n}}\\ &-\frac{2n^2\beta\mu^2}{(m+2n)(m+4n)}\Omega^{\frac{m+4\pi}{n}} - \frac{1}{15}\alpha\mu^3\Omega^5\\ &-\frac{3n^2\beta\mu^3}{2(m+3n)(m+5n)}\Omega^{\frac{m+5n}{n}} + \frac{n^2\beta\mu^2}{2(m+n)(m+4n)}\Omega^{\frac{m+4n}{n}}\\ &+\frac{n^2\beta\mu^3}{(m+2n)(m+5n)}\Omega^{\frac{n+5\pi}{n}} + \frac{\alpha\mu^4}{72}\Omega^6 + \frac{3n^2\beta\mu^4}{4(m+3n)(m+6n)}\Omega^{\frac{n+6\pi}{n}}\end{aligned}\right)$$

$$+\frac{1}{T}\left[\begin{aligned}&\alpha\left(\Omega T - \frac{T^2}{2} - \frac{\Omega^2}{2}\right)\\ &+\frac{n\beta}{m+n}\left(\Omega^{\frac{n+n}{n}}T - \frac{n}{m+2n}T^{\frac{\pi+2\pi}{2}} - \frac{(m+n)}{(m+2n)}\Omega^{\frac{m+2n}{n}}\right)\end{aligned}\right]$$

$$+A + C_D\left[\frac{\alpha\mu\Omega^2}{2} + \frac{n\beta\Omega^{\frac{\pi+2\pi}{n}}}{m+2n} + \frac{\alpha\mu^2\Omega^3}{2\times 3} + \frac{n\beta\mu^2\Omega^{\frac{\pi+3}{n}}}{2(m+3n)}\right] \tag{3.10}$$

$$\frac{\partial}{\partial\Omega}(\text{TSC}) = 0$$

In order to minimize the gross total inventory cost, we find the optimum values of variables T and τ. The optimum values of element expression T and τ may be obtained by equating the partial derivatives of total stock/inventory cost w.r.t. T and τ, respectively, to 0. For the optimum value of τ and T, we have

$$\frac{\partial(TSC)}{\partial\Omega} = 0 \tag{3.11}$$

$$\frac{\partial(TSC)}{\partial T} = 0 \tag{3.12}$$

The total minimum cost per unit time $\pi C(T,\Omega)$ is satisfied under the sufficient condition

$$\frac{\partial^2(TSC)}{\partial T^2} > 0 \tag{3.13}$$

$$\frac{\partial^2(TSC)}{\partial \Omega^2} > 0 \tag{3.14}$$

$$\frac{\partial^2(TSC)}{\partial T^2} * \frac{\partial^2(TSC)}{\partial \Omega^2} - \frac{\partial^2(TSC)}{\partial \Omega \partial T} > 0 \tag{3.15}$$

By solving equation (3.11), we determine the value of τ, and by solving equation (3.12), we determine the value of T. Then, by substituting these values into equation (3.10), we determine the lowest hourly cost for the values that meet the criteria outlined in equations (3.13), (3.14), and (3.15).

Therefore, the inventory model outlined above pertains to deteriorating items, with the demand rate increasing over time raised to the power of m/n. Additionally, constant parameters such as deterioration rate, holding, and ordering costs are taken into account. Shortages are permitted and completely accumulated.

3.3 ILLUSTRATIVE EXAMPLE

Let's consider a numerical expression and see whether the solution can be optimized for better results. Maple, a mathematical program, was used to solve the problem. To mathematically understand this model, the following parameters of an inventory-based system were used:

$\alpha = 13.790,$

$\beta = 10.031,$

$\Upsilon = 11.6,$

(Here, $\alpha, \beta,$ and Y are independent quantities used to represent the requirement or demand expression.)

[Cost based on holding per unit time is constant $\text{Hol}(t) = C_H$, where $Co_H \geq 0$.]

$C_{oH} = 27.781,$

Shortage cost per unit time, $Co_S = 15.209,$

$Os = 500, C_D = 10, m = 3, n = 4.$

The optimum shortage value is obtained from the above parameter:

$\tau = 0.2436509076, \text{Time} = 10.63800030.$

So, the total optimal cost of the model obtained is $TIc = 65594.10348$.

3.4 CONCLUSION

This chapter presents an inventory model that relies on time raised to the power of *m*/*n* for demand, along with a fixed reduction rate and consistent holding costs. The ideal expense was determined based on the given values and meeting the requirements. The proposed model has been subjected to mathematical and diagrammatic testing. According to this inventory model, demand varies as commodity prices increase regularly. This model provides a more precise representation of the environment. Future urban areas and businesses will greatly benefit from this idea.

BIBILIOGRAPHY

Amutha, R., & Chandrasekaran, D. E. (2012). An inventory model for deteriorating products with Weibull distribution deterioration, time-varying demand and partial backlogging. *International Journal of Scientific and Engineering Research*, *3*(10), 1–4.

Cheng, M., & Wang, G. (2009). A note on the inventory model for deteriorating items with trapezoidal type demand rate. *Computers & Industrial Engineering*, *56*(4), 1296–1300.

Ghosh, S., & Chaudhuri, K. (2004). An order-level inventory model for a deteriorating item with Weibull distribution deterioration, time-quadratic demand and shortages. *Advanced Modeling and Optimization*, *6*(1), 21–35.

Hou, K. L. (2006). An inventory model for deteriorating items with stock-dependent consumption rate and shortages under inflation and time discounting. *European Journal of Operational Research*, *168*(2), 463–474.

Jalan, A. K., & Chaudhuri, K. S. (1999). Structural properties of an inventory system with deterioration and trended demand. *International Journal of Systems Science*, *30*(6), 627–633.

Jalan, A. K., Giri, R. R., & Chaudhuri, K. S. (1996). EOQ model for items with Weibull distribution deterioration, shortages and trended demand. *International Journal of Systems Science*, *27*(9), 851–855.

Mandal, B., & Phaujdar, S. (2013). An inventory model for deteriorating items with stock-dependent consumption rate and shortage. *Journal of Information and Optimization Sciences*, *34*(2-3), 103–113.

Mandal, B. A., & Phaujdar, S. (1989). An inventory model for deteriorating items and stock-dependent consumption rate. *Journal of the Operational Research Society*, *40*(5), 483–488.

Maragatham, M., & Palani, R. (2017). An inventory model for deteriorating items with lead time price dependent demand and shortages. *Advances in Computational Sciences and Technology*, *10*(6), 1839–1847.

Mashud, A. H. M. (2020). An EOQ deteriorating inventory model with different types of demand and fully backlogged shortages. *International Journal of Logistics Systesssssms and Management*, *36*(1), 16–45.

Min, J., Zhou, Y. W., & Zhao, J. (2010). An inventory model for deteriorating items under stock-dependent demand and two-level trade credit. *Applied Mathematical Modelling*, *34*(11), 3273–3285.

Mishra, V. K. (2013). Deteriorating inventory model with quadratically time varying demand and partial backlogging. *WPOM-Working Papers on Operations Management*, *4*(2), 16–28.

Misra, R. B. (1975). Optimum production lot size model for a system with deteriorating inventory. *The International Journal of Production Research*, *13*(5), 495–505.

Rastogi, M., Singh, S., Kushwah, P & Tayal, S. (2017). An EOQ model with variable holding cost and partial backlogging under credit limit policy and cash discount. *Uncertain Supply Chain Management*, *5*(1), 27–42.

Roy, A. (2008). An inventory model for deteriorating items with price dependent demand and time varying holding cost. *Advanced Modeling and Optimization*, *10*(1), 25–37.

Shah, N. H., Chaudhari, U., & Jani, M. Y. (2016). An integrated production-inventory model with preservation technology investment for time-varying deteriorating item under time and price sensitive demand. *International Journal of Inventory Research*, *3*(1), 81–98.

Shah, N. H., & Naik, M. K. (2019). Coordinated production, ordering, shipment and pricing model for supplier-retailer inventory system under trade credit. *Operations Research and Decisions*, *29*(2), 55–76.

Shah, Y. K. (1977). An order-level lot-size inventory model for deteriorating items. *AIIE Transactions*, *9*(1), 108–112.

Shah, Y. K., & Jha, P. J. (1991). A single-period stochastic inventory model under the influence of marketing policies. *Journal of the Operational Research Society*, *42*(2), 173–176.

Sharma, A. K. (2019a). An overview and study on inventory model with deteriorating items. *International Journal of Scientific Research in Engineering and Management*, *3*(10), 1–5.

Sharma, A. K. (2019b). On some inventory model with deteriorating objects. *International Journal for Scientific Research and Development*, 7(8), 377–380.

Sharma, A. K., & Bansal, K. K. (2015). Analysis of deteriorating inventory model for trapezoidal type demand rate. *International Journal of Engineering Research & Management Technology*, 2(6), 151–159.

Sharma A. K., Bansal K. K. (2016). Analysis of inventory model with time dependent demand under fractional backlogging. *International Journal of Education and Science Research Review*, *3*(1), 103–108.

Singh, S. R., & Rathore, H. (2014, May). Optimal payment policy for deteriorating item with preservation technology investment under trade credit and inflation. In *International Conference on Recent Advances and Innovations in Engineering (ICRAIE-2014)* (pp. 1–6). IEEE, Jaipur, India.

Tripathy, C. K., & Mishra, U. (2015). An EOQ model with time dependent Weibull Debilitation and ramp type demand. *International Journal of Industrial Engineering Computations*, 2(2), 307–318.

Venkateswrlu R., & Mohan R (2014). An inventory model with quadratic demand, constant Debilitation and salvage value. *Research Journal of Mathematical and Statistical Sciences*, 2(1), 1–5.

Zhang, S., Cao, L., & Lu, Z. (2022). An EOQ inventory model for deteriorating items with controllable deterioration rate under stock-dependent demand rate and non-linear holding cost. *Journal of Industrial and Management Optimization*, *18*(6), 4231–4263.

Chapter 4

Analysis of particle swarm optimization-based inventory model with non-instantaneous deteriorating items

Ajender Kumar Malik, Priyanka Verma, Sharanjeet Dhawan, and Ather Aziz Raina

4.1 INTRODUCTION

Because of competition and uncertainty in the market, it is quite difficult to maintain stability in a business organization for every production and sales company. Holding and allocating an inventory of a product is one of the vital factors for relatively reasonably priced and the manipulate strolling of any business enterprise, be it a seven-star hotel, a printing press, a manufacturing business enterprise, or a hospital. Inventories form a substantial part of a corporation's total assets and require careful management. In regions where resources are highly constrained—such as India, Afghanistan, Sri Lanka, Bangladesh, Bhutan, Pakistan, Nepal, and Myanmar—the efficient and optimal utilization of available resources should be given top priority. Materials account for a top part of this standard charge, and inventories account for the essential part of the running capital. Therefore, the materials and inventory manipulation plays a crucial role in improving productivity.

The particle swarm optimization (PSO) has been broadly implemented for solving various real-time nonlinear optimization problems. Inventory problems concerning items subject to deterioration and variable demand are of considerable importance, as they contribute to the development of effective inventory control strategies and advance operations research. During the initial phase of a product's life cycle, the original quality of the item is maintained, as deterioration does not occur in this period. Ouyang et al. (2006) referred to this phenomenon as "non-immediate deterioration," and they evolved a non-instantaneous stock version for deteriorating products under permissible delay in payments. Manna and Chaudhuri (2006) and Zhou and Gu (2007) proposed the top-rated manufacturing stock coverage for the deteriorating products. Wu et al. (2006) proposed the optimal replenishment model which is considers for the non-instantaneous deterioration, stock-dependent demand, partial backlogging under the cost-effective ordering policies. Liao (2008) developed a financial order quantity (EOQ) stock version for non-instantaneous receipt using exponentially deteriorating products under the two-level alternate credits period. Li et al. (2010) cautioned against evaluation in deteriorating stock studies. Singh and Malik

 DOI: 10.1201/9781003521389-4

(2010, 2011) explored the inventory-structured demand stock version with storage capability for non-on-the-spot deteriorating items. Malik and Singh (2011) provided a comprehensive literature review of inventory control systems considering variable demand and soft computing techniques.

Gupta et al. (2013) proposed a most fulfilling ordering coverage with inventory-structured demand stock version for non-on-the-spot deteriorating gadgets. Sarkar and Sarkar (2013a) presented an advanced stock version for time-varying deterioration with inventory-structured call under partial backlogging. Sarkar and Sarkar (2013b) developed a financial amount version with probabilistic deterioration price in the manufacturing system. Singh et al. (2014) presented a review of inventory models under soft computing techniques. Some of the most important additional inventory papers that deal with the non-instantaneous deterioration in the inventory system are found in Singh et al. (2011), Sharma et al. (2013), Malik et al. (2016a, 2018, 2021), Vikram et al. (2016), Mathur et al. (2019). Ghoreishi et al. (2014) considered a most fulfilling pricing and ordering coverage for non-on-the-spot decaying products under the inflation. Sarkar et al. (2015) provided a stock version with nice development and back-order charge discount under controllable lead time. Malik et al. (2008, 2016b) explored the effect of time-varying demand rates in the inventory system. Yadav and Malik (2014) explored the optimization techniques with different models of project, queuing, inventory, and simulation techniques. Recently Tyagi et al. (2022) and Malik et al. (2022) proposed the effect of uncertainty in the inventory system. In a related study, Sharma et al. (2023) developed an inventory model centered on retailer–supplier trade-credit policies, examining the impact of credit terms on ordering behavior and inventory costs to support improved inventory management decisions. Jayaswal et al. (2024) proposed a sustainable ordering policy for deteriorating items by incorporating trade-credit financing and the learning effect, highlighting their influence on operational efficiency and optimal ordering decisions.

The application of PSO for solving the inventory system in production and sales has become a new trend in recent years. Various methods of optimization can be used for solving industrial problems. Poli et al. (2007) provided a general overview of PSO guidelines. PSO was originally proposed by Kennedy and Eberhart in 1995 and subsequently enhanced by Shi and Eberhart in 1998. The technique is a population-based stochastic optimization method in which a swarm of particles explores the search space, with each particle adjusting its position based on its own experience and the collective knowledge of the swarm. Mousavi et al. (2017) and Kumar et al. (2022) reviewed PSO guidelines to solve containment gap and inventory management problems in a chain network.

Due to the inherent nonlinearity of the proposed problem, suitable optimization strategies, including PSO techniques, are adopted. We try to interpret the numerical results to understand their meaning. Most fashion researchers no longer consider the non-spatial and temporal decay phenomena at the same time, so we include them in our version. A numerical instance with graphical representations is furnished to have a look at the impact of adjustments in the related parameters on the most fulfilling answer.

4.2 NOTATIONS AND ASSUMPTIONS

For the formulation of the present inventory model, the notations and assumptions listed below in the table are employed:

$D = d_1 - d_2 t$	*Demand for the product*
Co	Ordering cost per order
C_h	Holding cost per unit time per item
C_p	Purchasing cost per item
L	Maximum inventory level
C_d	Deteriorating cost per item
$\theta(t)$	Time-dependent deteriorating rate, $\theta(t) = \frac{1}{1+R-t}$, where $0 \le \theta(t) \le 1$,
R	Maximum lifetime for the product
t_1	No deterioration in this time period
T	Total cycle
TCF	Total inventory cost function

Assumptions

1. Demand rate of the product is a linear decreasing function of time.
2. The lead time is zero or negligible.
3. The shortages are not approved in this system.

4.3 MATHEMATICAL FORMULATION

In this section, the inventory level (I_1) decreases due to the demand rate in the time interval $[0, t_1]$ and no deterioration (Figure 4.1). After time t_1, both the demand and deterioration start immediately and stop when the inventory diminishes to level zero.

The inventory levels of the developing system using the differential equations in the time interval $[0, T]$ can be established:

$$\frac{dI_1(t)}{dt} = -(d_1 - d_2 t) \quad 0 \le t \le t_1 \tag{4.1}$$

$$\frac{dI_2(t)}{dt} + \frac{1}{1+R-t} I_2(t) = -(d_1 - d_2 t) \quad t_1 \le t \le T \tag{4.2}$$

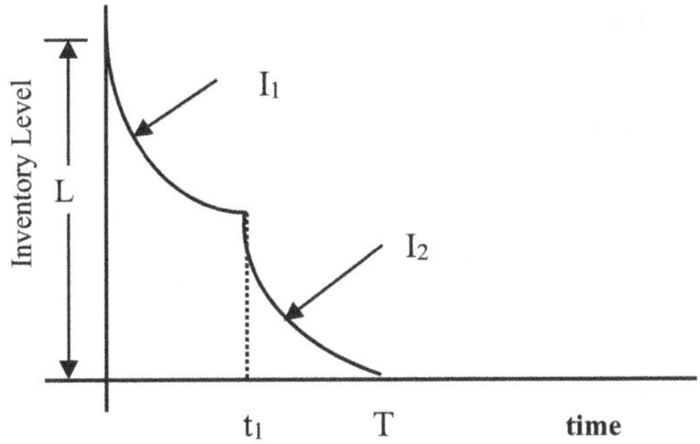

Figure 4.1 Inventory model.

boundary conditions $I_1(0) = L$, $I_1(T) = 0$, respectively. Using Equations (4.1) and (4.2), we get

$$I_1(t) = L - d_1 t + d_2 \frac{t^2}{2}, \quad 0 \le t \le t_1 \tag{4.3}$$

$$I_2(t) = (1 + R - t)\left[d_3 \log\left(\frac{1 + R - t}{1 + R - T}\right) + d_2(T - t)\right], \qquad t_1 \le t \le T \tag{4.4}$$

where $d_3 = d_1 - d_2(1 + R)$

Due to the continuity of $I(t)$ at $t = t_1$, from equations (4.3) and (4.4), we have

$$\Rightarrow \quad L = d_1 t_1 - d_2 \frac{t_1^2}{2} + \left(d_3 \log\left(\frac{1 + R - t_1}{1 + R - T}\right) + d_2(T - t_1)\right) \tag{4.5}$$

The optimum inventory cost per cycle contains the following values:

Ordering cost per cycle is $OC = C_o$ (4.6)

Holding cost per cycle

$$HC = C_h\left(\int_0^{t_1} I_1(t)dt + \int_{t_1}^{T} I_2(t)dt\right)$$

$$= C_h \left[\begin{array}{l} Lt_1 - \frac{d_1}{2}t_1^2 + \frac{d_2}{6}t_1^3 + (1 + R)\left\{d_3(T - t_1) + \frac{d_2}{2}(T - t_1)^2\right\} \\ + \frac{d_3}{4}(T^2 - t_1^2) + \frac{d_3(1 + R)}{2}(T - t_1) \\ - \frac{d_2}{6}(T^3 + 2t_1^3 - 3Tt_1^2) \\ + d_3 \log\left(\frac{1 + R - t_1}{1 + R - T}\right)\left\{\frac{(1 + R)^2}{2} - t_1(1 + R) + \frac{t_1^2}{2}\right\} \end{array}\right] \tag{4.7}$$

Deterioration cost per cycle

$$DC = C_d \int_{t_1}^{T} \theta(t)\ I_2(t)dt$$
$$= C_d\left[d_3(T-t_1)+d_3(1+R-t_1)\log\left(\frac{1+R-t_1}{1+R-T}\right)+\frac{d_2}{2}(T-t_1)^2\right] \quad (4.8)$$

Purchasing cost per cycle

$$PC = C_p \times L = C_p\left[\begin{array}{l} d_1t_1 - d_2\frac{t_1^2}{2} \\ +(1+R-t_1)\left(d_3\log\left(\frac{1+R-t_1}{1+R-T}\right)+d_2(T-t_1)\right)\end{array}\right] \quad (4.9)$$

Thus, the optimum total cost function (TCF) per cycle per unit time is

$$TCF = \frac{1}{T}[OC + HC + DC + PC] \quad (4.10)$$

To determine the optimal cycle time T^*, we minimize the total inventory cost of the function. To minimize the TCF, we have

$$\frac{dTCF}{dT} = 0 \quad \text{and} \quad \frac{d^2TCF}{dT^2} > 0 \quad (4.11)$$

The inventory cost function is highly nonlinear in nature.

4.4 PARTICLE SWARM OPTIMIZATION

PSO technique, an especially search technique, was established on the behavior of flocking of birds. Eberhart and Kennedy developed this algorithm. The proposed strategies assumed the modeling of stock manipulation with PSO strategies under variable deterioration quotes and income sales prices, offering a brand-new measurement for the betterment of the producing and commercial areas by Shekhar et al. (2020, 2021). Generally, managers depend on mathematical calculation and computer structures developed by mathematicians, statisticians, and commercial engineers to deal with the state of affairs of problem-solving in stock manipulate and management of the system. There are as many fashions viable as there are businesses because each has a distinctive price shape and constraints (Figure 4.2).

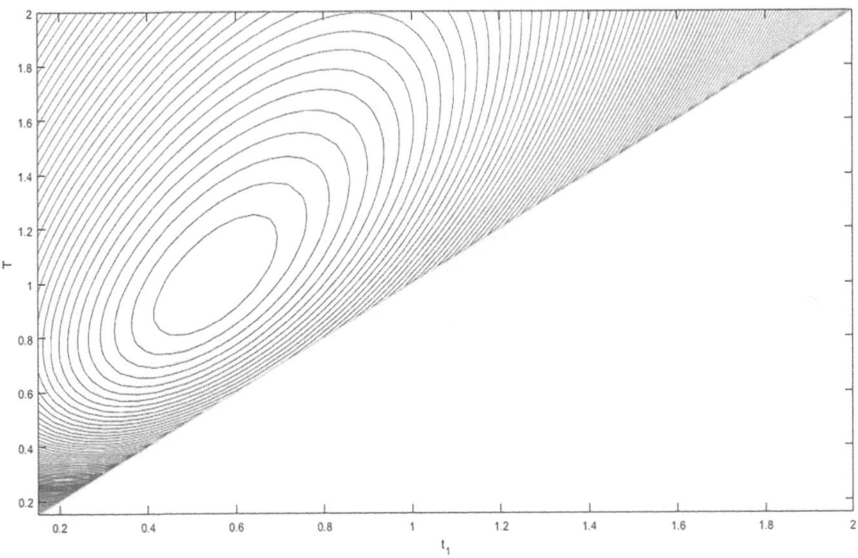

Figure 4.2 Graphical representation of PSO.

4.5 NUMERICAL EXAMPLE

Due to incorrectness and uncertainty of the parameters, it is very difficult and risky for a business manager of a production firm to make a perfect model with exact data for their future work. However, we take the following values of the parameters used: $C_o = 100$, $C_p = 10$, $C_h = 0.25$, $C_d = 5$, $R = 25$, and $d_1 = 1000$, $d_2 = 10$. Using the PSO techniques, the optimal ordering cycle $T^* = 1.4241$ and the corresponding $TPF^* = 1124.21$ and $L^* = 104.72$.

C_o	C_p	C_h	C_d	t_1^*	T^*	L^*	TCF
100	10	0.25	5	0.5448	1.4241	104.72	1124.21
200	10	0.25	5	0.9048	1.4451	147.21	1252.38
300	10	0.25	5	0.9132	1.7998	115.71	1430.41
100	5	0.25	5	0.6611	1.9951	103.93	541.32
100	15	0.25	5	0.4129	3.4156	108.41	1717.99
100	25	0.25	5	0.3929	4.4129	106.52	1162.81
100	10	0.4	5	0.4446	0.4142	116.41	1599.71
100	10	0.6	5	0.4741	0.8334	102.61	1123.84
100	10	0.8	5	0.4859	0.9324	111.62	1253.84
100	10	0.25	2	0.7241	1.8841	104.40	1254.87
100	10	0.25	3	0.7346	1.9921	170.74	1354.64
100	10	0.25	4	0.7447	1.9762	100.21	1359.81

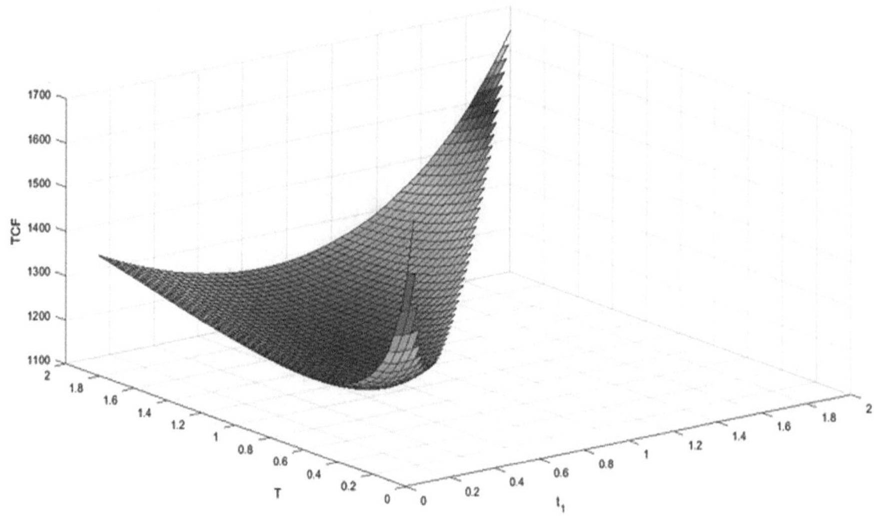

Figure 4.3 Total inventory cost (TCF) w.r.t. T and t_1.

The following graph (Figure 4.3) shows that the relation between the total inventory cost TCF and the time duration t_1 and T.

4.6 CONCLUSION

In most of the higher inventory models, the researchers and authors indicate a fixed rate of deterioration. However, in general terms, the maximum number of goods/items is transferred to the pot since their maximum lifespan is about to expire. However, according to the authors, a form of viewing time as a function of diverse weather decline, with the assumption of a non-instantaneous commodity, has not yet been proposed. This audited build and evaluation is within the well-known agenda and can be extended to include optimization strategies, and PSO algorithms helps professionals with seasonal goods, including fruits, vegetables, packaged foods, medicines, luxury goods, and more. The concept of this optimized model of the study work can be applied in various business industries, sectors, production companies, and sales markets. Finally, a numerical evaluation of the cost/benefit of the standard stock is described under multiple constraints to arrive at a consistent result. For further investigation, the study may consider inventory versions with changing demand, variable cost of ownership, inflation, probabilistic demand, partial ordering, reliability, and uncertainty environments.

REFERENCES

Ghoreishi, M., Mirzazadseh, A., & Weber, G. W. (2014). Optimal pricing and ordering policy for non-instantaneous deteriorating items under inflation and customer returns. *Optimization*, *63*(12), 1785–1804.

Gupta, K. K., Sharma, A., Singh, P. R., & Malik, A. K. (2013). Optimal ordering policy for stock-dependent demand inventory model with non-instantaneous deteriorating items. *International Journal of Soft Computing and Engineering*, *3*, 279–281.

Jayaswal, M. K., Malik, A. K., & Kumar, V. (2024). A sustainable ordering policies model with trade-credit for deteriorating items under the learning effect. *Journal of Management Analytics*, *11*(4), 631–658.

Kennedy, J., & Eberhart, R. (1995). Particle swarm optimization. *Paper presented at IEEE International Conference on Neural Networks*, Vol. 4, 1942–1948. Perth, WA, Australia: IEEE.

Kumar, A., Boualem, M., & Bouchentouf, A. A. (2022). Optimal analysis of machine interference problem with standby, random switching failure, vacation interruption and synchronized reneging. In A. Goel, A. Chauhan, A. K. Malik (eds), *Applications of Advanced Optimization Techniques in Industrial Engineering* (pp. 155–168). CRC Press.

Li, R., Lan, H., & Mawhinney, J. R. (2010). A review on deteriorating inventory study. *Journal of Service Science and Management*, *3*(1), 117–129.

Liao, J. J. (2008). An EOQ model with non-instantaneous receipt and exponentially deteriorating items under two-level trade credit. *International Journal of Production Economics*, *113*(2), 852–861.

Malik, A. K., Bano, T., Pandey, M. K., & Bansal, K. K. (2021). Inflation based Inventory model among non-instantaneous decaying items with linear demand rate. *Annals of Optimization Theory and Practice*, e136652.

Malik, A. K., Sharma, M., Tyagi, T., Kumar, S., Naik, P. J., & Kumar, P. (2022). Effect of uncertainty in optimal inventory policy for manufacturing products. *International Journal of Intelligent Systems and Applications in Engineering*, *10*(1s), 102–110.

Malik, A. K., Shekhar, C., Vashisth, V., Chaudhary, A. K., & Singh, S. R. (2016a, March). Sensitivity analysis of an inventory model with non-instantaneous and time-varying deteriorating Items. In *AIP Conference Proceedings*, Vol. 1715, No. 1, p. 020059. Pilani: AIP Publishing LLC.

Malik, A. K., Singh, P. R., Tomar, A., Kumar, S., & Yadav, S. K. (2016b, March). Analysis of an inventory model for both linearly decreasing demand and holding cost. In *AIP Conference Proceedings*, Vol. 1715, No. 1, p. 020063. Pilani: AIP Publishing LLC.

Malik, A. K., Singh, S. R., & Gupta, C. B. (2008). An inventory model for deteriorating items under FIFO dispatching policy with two warehouse and time dependent demand. *Ganita Sandesh*, *22*(1), 47–62.

Malik, A. K., & Singh, Y. (2011). An inventory model for deteriorating items with soft computing techniques and variable demand. *International Journal of Soft Computing and Engineering*, *1*(5), 317–321.

Malik, A. K., Vedi, P., & Kumar, S. (2018). An inventory model with time-varying demand for non-instantaneous deteriorating items with maximum life time. *International Journal of Applied Engineering Research*, *13*(9), 7162–7167.

Manna, S. K., & Chaudhuri, K. S. (2006). An EOQ model with ramp type demand rate, time dependent deterioration rate, unit production cost and shortages. *European Journal of Operational Research*, *171*(2), 557–566.

Mathur, P., Malik, A. K., & Kumar, S. (2019, August). An inventory model with variable demand for non-instantaneous deteriorating products under the permissible delay in payments. In *IOP Conference Series: Materials Science and Engineering*, Vol. 594, No. 1, p. 012042. Jaipur: IOP Publishing.

Mousavi, S. M., Bahreininejad, A., Musa, S. N., & Yusof, F. (2017). A modified particle swarm optimization for solving the integrated location and inventory control problems in a two-echelon supply chain network. *Journal of Intelligent Manufacturing*, *28*(1), 191–206.

Ouyang, L. Y., Wu, K. S., & Yang, C. T. (2006). A study on an inventory model for non-instantaneous deteriorating items with permissible delay in payments. *Computers & Industrial Engineering*, *51*(4), 637–651.

Poli, R., Kennedy, J., & Blackwell, T. (2007). Particle swarm optimization. *Swarm Intelligence*, *1*(1), 33–57.

Sarkar, B., Mandal, B., & Sarkar, S. (2015). Quality improvement and backorder price discount under controllable lead time in an inventory model. *Journal of Manufacturing Systems*, *35*, 26–36.

Sarkar, B., & Sarkar, S. (2013a). An improved inventory model with partial backlogging, time varying deterioration and stock-dependent demand. *Economic Modelling*, *30*, 924–932.

Sarkar, M., & Sarkar, B. (2013b). An economic manufacturing quantity model with probabilistic deterioration in a production system. *Economic Modelling*, *31*, 245–252.

Sharma, A., Gupta, K. K., & Malik, A. K. (2013). Non-instantaneous deterioration inventory model with inflation and stock-dependent demand. *International Journal of Computer Applications*, *67*(25), 6–9.

Sharma, A., Singh, C., Malik, A. K., & Modibbo, U. M. (2023). Inventory model for retailer-supplier's trade credit policy. *International Journal of Applied Optimization Studies*, *3*(1), 1–23.

Shekhar, C., Varshney, S., & Kumar, A. (2020). Optimal control of a service system with emergency vacation using bat algorithm. *Journal of Computational and Applied Mathematics*, *364*, 112332.

Shekhar, C., Varshney, S., & Kumar, A. (2021). Matrix-geometric solution of multi-server queueing systems with Bernoulli scheduled modified vacation and retention of reneged customers: A meta-heuristic approach. *Quality Technology & Quantitative Management*, *18*(1), 39–66.

Shi, Y., & Eberhart, R. (1998, May). A modified particle swarm optimizer. In *Evolutionary Computation Proceedings*, Vol. 890, pp. 69–73.

Singh, S. R., & Malik, A. K. (2010). Optimal ordering policy with linear deterioration, exponential demand and two storage capacity. *International Journal of Mathematical Sciences*, *9*(3–4), 513–528.

Singh, S. R., & Malik, A. K. (2011). An inventory model with stock-dependent demand with two storages capacity for non-instantaneous deteriorating items. *International Journal of Mathematical Sciences and Applications*, *1*(3), 1255–1259.

Singh, S. R., Malik, A. K., & Gupta, S. K. (2011). Two warehouses inventory model for non-instantaneous deteriorating items with stock-dependent demand. *International Transactions in Applied Sciences*, *3*(4), 911–920.

Singh, Y., Arya, K., & Malik, A. K. (2014). Inventory control with soft computing techniques. *International Journal of Innovative Technology and Exploring Engineering*, *3*(8), 80–82.

Tyagi, T., Kumar, S., Malik, A. K., & Vashisth, V. (2022). A novel neuro-optimization technique for inventory models in manufacturing sectors. *Journal of Computational and Cognitive Engineering*, 2(3), 204–209.

Vikram, V., Ajay, T., Chandra, S., & Malik, A. K. (2016). A trade credit inventory model with multivariate demand for non-instantaneous decaying products. *Indian Journal of Science and Technology*, *9*(15), 1–6.

Wu, K. S., Ouyang, L. Y., & Yang, C. T. (2006). An optimal replenishment policy for non-instantaneous deteriorating items with stock-dependent demand and partial backlogging. *International Journal of Production Economics*, *101*(2), 369–384.

Yadav, S. R., & Malik, A. K. (2014). *Operations Research*. Oxford University Press.

Zhou, S. Y., & Gu, F. W. (2007). An optimal production-inventory policy for deteriorating items. *Logistics Technology*, *26*(4), 43–46.

Chapter 5

A novel approach to mathematical modeling on sustainability of mixing problems with renewable energy

Shiv Raj Singh and Dipti Singh

5.1 INTRODUCTION

Energy demand is increasing day by day due to technological development, but the amount of available energy is not enough. The price of energy per unit increases daily. A renewable energy source can be used as an alternative to traditional energy sources at the moment. Waste can be converted into renewable energy. A major benefit of this system is its ability to reduce environmental pollution through the utilization of waste. It is possible to use renewable energy in many different ways. Nowadays, environmental issues are considered in the inventory models to reduce greenhouse gas emissions all over the world. Investment in high-quality green technology can reduce environmental emissions. Managers seek to reduce the environmental effects of the manufacturing system while striving to achieve environmentally friendly production. Under different carbon emission regulations, this study presents a quantitative economic model that takes green technology into account. The importance of studying the mixing problem of liquid from a global perspective is gaining momentum. In different areas of business management, mixing is a problem mostly in liquid products, like chemical products, pesticides, medicines, oils, mineral water, and detergents. So, the mathematical modeling in the mixing problems is very much necessary to develop an inventory model for such types of liquid products, along with sustainability.

In this direction, some research on solid products has been conducted in various disciplines like engineering, computer science, and mathematics. But there is a vacuum for liquid products toward sustainability. Very little research has been done on mixing liquid products, although it is a very important and realistic situation. To fill this vacuum, we develop a mathematical model for mixing liquid products toward sustainability with the help of renewable energy.

 DOI: 10.1201/9781003521389-5

5.2 LITERATURE REVIEW

In recent years, renewable energy has played an increasingly important role in sustainable development. Many scientists and researchers are working in this area to reduce the greenhouse gas (GHG) emissions. According to Kök et al. (2018), peak-pricing produces better investment results than flat-pricing for renewable energy investments. In their paper, they discussed the investment in renewable energy. Datta (2017) examined the effects of green technology investments under the cap tax system, which optimizes renewable energy investments in transportation. According to Sarkar et al. (2018a), the cost of non-renewable energy is considerable for a production system based on a traditional production model. By using renewable energy, the cost can be reduced. In Sarkar et al. (2018b), failure rates in a smart production system for defective products were discussed. Additionally, their study concluded that flexible and smart production systems require renewable energy. A two-level trade credit policy was incorporated by Singh et al. (2021) to examine the impact of energy use and carbon emissions on supply chain management. Sarkar and Chung (2021) developed an inventory model on "Effect of Renewable Energy to Reduce Carbon Emissions under a Flexible Production System: A Step Toward Sustainability". Jaggi et al. (2022) created a sustainable inventory model that incorporates renewable energy into the production process. Numerous studies have focused on investing in green technology to decrease carbon emissions and promote environmental sustainability. Recently, Hasan et al. (2021) formulated an inventory model in which the carbon emission effect is reduced by investment in green technology. Sepehri et al. (2021) developed a combined pricing and inventory model for perishable items with a defined maximum lifetime and manageable carbon emissions, considering permissible payment delays. In a fuzzy environment, Sharma et al. (2021) developed a reverse logistics inventory model with multiple productions and remanufacturing batches. Very little research is done on the inventory model with the mixing liquid problem. Rajabi et al. (2020) developed a model for "Emissions of volatile organic compounds from crude oil processing – Global emission inventory and environmental release". Rahman et al. (2021) developed an inventory model in which the mixing process of liquid is mathematically given by a simultaneous differential equation. It was developed by Vitale et al. (2022) that the material flow is modeled and optimized in the wood pellet supply chain.

In this paper, we developed a mathematical model in which liquid products are taken and renewable energy is used in the flexible production system and storage process. Also, green technology investment is used to reduce carbon emissions.

5.3 NOTATIONS AND ASSUMPTIONS

5.3.1 Notations

Notations are given below in Table 5.1.

Table 5.1 Notations used in the proposed paper are shown

Notation	*Explanations*
$\eta(t)$	Amount of liquid in container I at time t (%)
$\delta(t)$	Amount of liquid in container II at time t (%)
$I(t)$	Mixed liquid stock level
ϕ_1	Volume in liters of the container I
ϕ_2	Volume in liters of the container II
α_1	Quantity of liquid entering container I (unit time) with concentration k
α_2	Container I liquid outflow rate (L/unit time)
β_1	The rate at which liquid enters container I from container II (in liters per unit of time)
β_2	Liquid 2 egress rate from container II (liter per unit time)
γ	Liquid concentration in the container at the beginning II (%)
k	Liquid concentration in an external supply (%)
$D(p)$	Customer demand
$P(t)$	Rate of produce items
θ	Rate of deterioration
G	Investment in green technology
p	Selling price
C_o	Setup cost ($/ setup)
C_d	Cost per unit of degradation over time ($/ Setup)
h	Storage cost per unit per unit time
h'	Renewable energy cost for storing the product
ω	Tool/die cost per unit
ω'	Renewable energy cost during production
γ	Product development cost per unit
γ'	Energy cost associated with a sustainable production system
ψ	Material cost per unit
t_1	Time at which production stops
T	Duration of a cycle (time unit)
$TP(t_1, p, T)$	Profitability average ($/unit time)

5.3.2 Assumptions

Assumptions are given as below:

1. Investment is allowed in this model.
2. Renewable energy is used in the production process and the storage process.
3. Investment in green technology is taken into consideration.
4. Time horizon is infinite.
5. This study addresses a mixture composed of three distinct concentrations of liquid (Rahman et al. 2021).
6. Container I has a capacity filled with liquid starting with a concentration of zero, while container II is filled with liquid starting with a concentration of k.
7. "As a first step, liquid with concentration γ is sent to container I at the rate α_1; then the mixture of liquid is sent to container II at the rate α_2; then the liquid is returned to container I at the rate β_1 from container II. The process continues to achieve the best desirable mixture, and once it has been reached, it is used as a production liquid from container II at the rate β_2" (Rahman et al. 2021).
8. The mixed liquid production rate $P(t)$ is directly proportional to the concentration level of the mixed liquid, $y(t)$. The mathematically represented by $P(t) = \dfrac{\beta_2 \delta(t)}{\phi_2}$.
9. The demand depends on selling price

   $D = (a - bp), a, b, p > 0 \text{ s.t } p < \dfrac{a}{b}$
10. Lead time is zero.

5.4 MATHEMATICAL MODELING

This model is developed for a manufacturer of liquid products who faces many problems in mixing liquids. The manufacturer starts production at time $t = 0$. As production starts, the demand for customer also starts to fulfill. At time $t = t_1$, the inventory level reaches the maximum level. At this, we stop the production. After that, the manufacturer's inventory level decreases due to the combined effect of deterioration and demand. Mathematical formulation for mixing problem is given by Rahman et al. (2021). They give

$$\eta(t) = e^{\left(\frac{-k_1 t}{2}\right)} \left\{ \phi_1 e^{k_4 t} + \phi_2 e^{-k_4 t} \right\} + \frac{k_3}{k_2} \tag{5.1}$$

$$\delta(t) = \frac{\beta_2}{\phi_2}[c_1\left(k_4 - \frac{k_1}{2} + \frac{\alpha_2}{\phi_1}\right)e^{\left(k_4 - \frac{k_1}{2}\right)t} + c_2\left(-k_4 - \frac{k_1}{2} + \frac{\alpha_2}{\phi_1}\right)$$
$$e^{\left(-k_4 - \frac{k_1}{2}\right)t}] - \frac{x\alpha_1\phi_2}{\eta} + \frac{\alpha_2\phi_2 k_3}{\beta_1\phi_1 k_2} \tag{5.2}$$

where

$$C_1 = \frac{-k_3}{2k_2} + \frac{2}{2k_4}\left[k\eta + x\alpha_1 + \frac{k_3}{k_2}\left(\frac{\alpha_2}{\phi_1} - \frac{k_2}{2}\right) - \frac{\alpha_2 k_3}{\phi_1 k_2}\right]$$

and

$$C_2 = \frac{-k_3}{2k_2} - \frac{1}{2k_4}\left[k\eta + x\alpha_1 + \frac{k_3}{k_2}\left(\frac{\alpha_2}{\phi_1} - \frac{k_2}{2}\right) - \frac{\alpha_2 k_3}{\phi_1 k_2}\right]$$

$$k_1 = \frac{\alpha_2}{\phi_1} + \frac{(\eta + \beta_2)}{\phi_2}, \quad k_2 = \frac{\alpha_2\beta_2}{\phi_1\phi_2}, \quad k_3 = \frac{\alpha_1(\beta_1 + \beta_2)}{\phi_1}$$

$$k_4 = \frac{\sqrt{\left(\frac{\alpha_2}{\phi_1} + \frac{(\eta+\beta_2)}{\phi_2}\right)^2 - \frac{4\alpha_2\beta_2}{\phi_1\phi_2}}}{2}$$

The inventory level shown in Figure 5.1 is given by

$$I_1(t) = \left(\frac{\alpha_2\beta_2 k_3}{\eta\theta k_2\phi_1} - \frac{D}{\theta} - \frac{x\phi_1\beta_2}{\eta}\right)\left(1 - e^{-\theta t}\right) + \frac{c_1}{\left(k_4 - \frac{k_1}{2} + \theta\right)}$$
$$\left\{\frac{\left(k_4 - \frac{k_1}{2}\right)\beta_2}{\eta} - \frac{\alpha_2\beta_2}{\eta\phi_1}\right\}\left(e^{\left(k_4 - \frac{k_1}{2}\right)t} - e^{-\theta t}\right) + \frac{c_2}{\left(k_4 - \frac{k_1}{2} - \theta\right)} \tag{5.3}$$
$$\left\{\frac{\left(k_4 + \frac{k_1}{2}\right)\beta_2}{\eta} - \frac{\alpha_2\beta_2}{\eta\phi_1}\right\}\left(e^{-\left(k_4 - \frac{k_1}{2}\right)t} - e^{-\theta t}\right), 0 \le t \le t_1$$

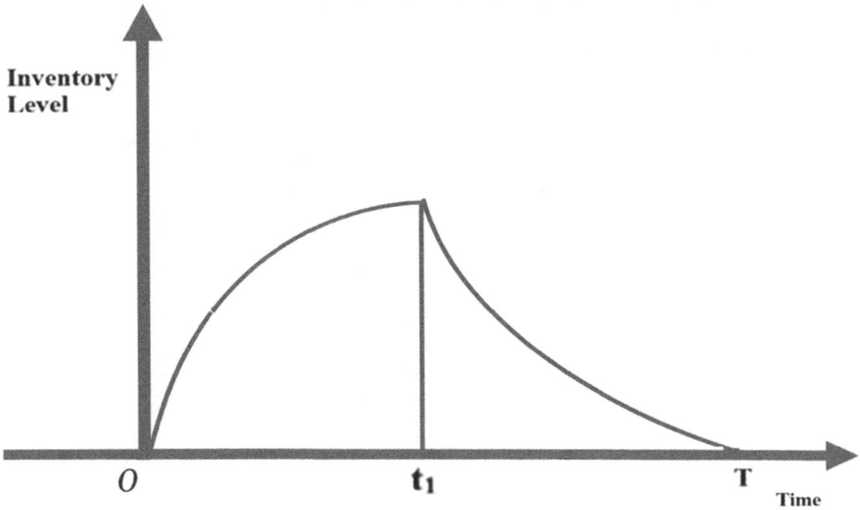

Figure 5.1 Graphical representation of different inventory levels.

and

$$I_2(t) = \frac{(a-bp)}{\theta}\left[e^{\theta(T-t)} - 1\right], t_1 \le t \le T \tag{5.4}$$

Various inventory costs are given by

1. Sales Revenue

$$S.R. = p\int_0^T D dt$$

$$S.R. = pDT = p(a-bp)T \tag{5.5}$$

2. Setup Cost

$$S.C. = C_0 \tag{5.6}$$

3. Production costs include raw material costs, dependent costs, and tooling and die costs. In addition to raw materials, energy costs have a significant impact on tooling and die costs as well.

The flexible cost due to production is obtained as

The production cost per unit

$$PC = \left(\psi + \frac{(\gamma + \gamma')}{\int_0^{t_1} P(t)\,dt} + (\omega + \omega') \int_0^{t_1} P(t)\,dt \right)$$

$$\begin{aligned}
PC = (\psi + {} & \frac{(\gamma + \gamma')}{\left\{ \frac{\beta_2}{\eta} \left[\frac{c_1\left(k_4 - \frac{k_1}{2} + \frac{\alpha_2}{\phi_1}\right)\left(e^{\left(k_4 - \frac{k_1}{2}\right)t_1} - 1\right)}{\left(k_4 - \frac{k_1}{2}\right)} + \frac{c_2\left(-k_4 - \frac{k_1}{2} + \frac{\alpha_2}{\phi_1}\right)\left(1 - e^{\left(-k_4 - \frac{k_1}{2}\right)t_1}\right)}{\left(k_4 - \frac{k_1}{2}\right)} \right] + \frac{\beta_2}{\eta}\left(\frac{\alpha_2 k_3}{\phi_1 k_2} - x\alpha_1\right)t_1 \right\}} + \\
& (\omega + \omega') \left\{ \frac{\beta_2}{\eta} \left[\frac{c_1\left(k_4 - \frac{k_1}{2} + \frac{\alpha_2}{\phi_1}\right)\left(e^{\left(k_4 - \frac{k_1}{2}\right)t_1} - 1\right)}{\left(k_4 - \frac{k_1}{2}\right)} + \frac{c_2\left(-k_4 - \frac{k_1}{2} + \frac{\alpha_2}{\phi_1}\right)\left(1 - e^{\left(-k_4 - \frac{k_1}{2}\right)t_1}\right)}{\left(k_4 - \frac{k_1}{2}\right)} \right] + \frac{\beta_2}{\eta}\left(\frac{\alpha_2 k_3}{\phi_1 k_2} - x\alpha_1\right)t_1 \right\}
\end{aligned} \tag{5.7}$$

4. **Holding Cost**

$$H.C. = (h+h')\left[\int_0^{t_1} I_1(T)\,dt + \int_{t_1}^{T} I_2(t)\,dt\right]$$

$$H.C. = (h+h')\left[\left(\frac{\alpha_2\beta_2 k_3}{\eta k_2\phi_1} - \frac{(a-bp)}{\theta} - \frac{x\phi_1\beta_2}{\eta\theta}\right)\left(t_1 - \frac{\left(1-e^{-\theta t_1}\right)}{\theta}\right) + \right.$$

$$\frac{c_1\left\{\frac{\left(k_4-\frac{k_1}{2}\right)\beta_2}{\eta} + \frac{\alpha_2\beta_2}{\phi_1\eta}\right\}\left[\frac{\left(e^{\left(k_4-\frac{k_1}{2}\right)t_1}-1\right)}{\left(k_4-\frac{k_1}{2}\right)} - \frac{\left(1-e^{\theta t_1}\right)}{\theta}\right]}{\left(k_4-\frac{k_1}{2}+\theta\right)} + \tag{5.8}$$

$$\frac{c_2\left\{\frac{\left(k_4+\frac{k_1}{2}\right)\beta_2}{\eta} - \frac{\alpha_2\beta_2}{\phi_1\eta}\right\}\left[\frac{\left(1-e^{\left(k_4-\frac{k_1}{2}\right)t_1}\right)}{\left(k_4-\frac{k_1}{2}\right)} - \frac{\left(1-e^{\theta t_1}\right)}{\theta}\right]}{\left(k_4+\frac{k_1}{2}-\theta\right)} +$$

$$\left.\frac{(a-bp)}{\theta^2}\left\{e^{\theta(T-t_1)}-1\right\}\right]$$

5. **Investment in Green Technology**

$$G.T. = GT \tag{5.9}$$

6. **Deterioration Cost**

$$D.C. = C_d\theta\left[\left(\frac{\alpha_2\beta_2k_3}{\eta k_2\phi_1} - \frac{(a-bp)}{\theta} - \frac{x\phi_1\beta_2}{\eta\theta}\right)\left(t_1 - \frac{\left(1-e^{-\theta t_1}\right)}{\theta}\right) + \frac{c_1\left\{\frac{\left(k_4 - \frac{k_1}{2}\right)\beta_2}{\eta} + \frac{\alpha_2\beta_2}{\phi_1\eta}\right\}\left[\frac{\left(e^{\left(k_4-\frac{k_1}{2}\right)t_1} - 1\right)}{\left(k_4 - \frac{k_1}{2}\right)} - \frac{\left(1-e^{\theta t_1}\right)}{\theta}\right]}{\left(k_4 - \frac{k_1}{2} + \theta\right)} + \frac{c_2\left\{\frac{\left(k_4 + \frac{k_1}{2}\right)\beta_2}{\eta} - \frac{\alpha_2\beta_2}{\phi_1\eta}\right\}\left[\frac{\left(1 - e^{\left(k_4-\frac{k_1}{2}\right)t_1}\right)}{\left(k_4 - \frac{k_1}{2}\right)} - \frac{\left(1-e^{\theta t_1}\right)}{\theta}\right]}{\left(k_4 + \frac{k_1}{2} - \theta\right)} + \frac{(a-bp)}{\theta^2}\left\{e^{\theta(T-t_1)} - 1\right\}\right] \quad (5.10)$$

So, the total profit is given by

$$TP = \frac{1}{T}\begin{bmatrix}\text{sales revenue} - \text{flexible production cost} - \text{setup cost} - \text{holding cost} \\ - \text{investment in green technology} - \text{deterioration cost}\end{bmatrix}$$

5.5 NUMERICAL ILLUSTRATION

Consider a situation in which all the parameters are given as follows:

$a = 1000$, $b = 0.3$, $\phi_1 = 690$, $\phi_2 = 700$, $\theta = 0.2$, $h = 30$, $x = 0.9$, $\alpha_1 = 325$, $\alpha_2 = 690$, $\beta_1 = 365$, $\beta_2 = 325$, $k = 0.5$, $h' = 1$, $C_0 = 450$, $G = 360$, $\gamma = 150$, $\gamma' = 100$, $C_d = 20\$/\text{unit}$, $\eta = 365$, $\psi = 20\$/\text{unit}$, $\omega = 0.8\$/\text{unit}$, $\omega' = 0.5$.

We get the optimal solution:

Total profit $TP = 845159.7\$$, $t_1 = 1.2$ years, $p = 1654.491$ Rs, $T = 1.54$ years.

5.6 CONCAVITY

The concavity of the total profit (TP), time (t_1), and cycle length (T) is shown in Figure 5.2. The concavity between total profit (TP), cycle length ("T"), and selling price ("p") is shown in Figure 5.3.

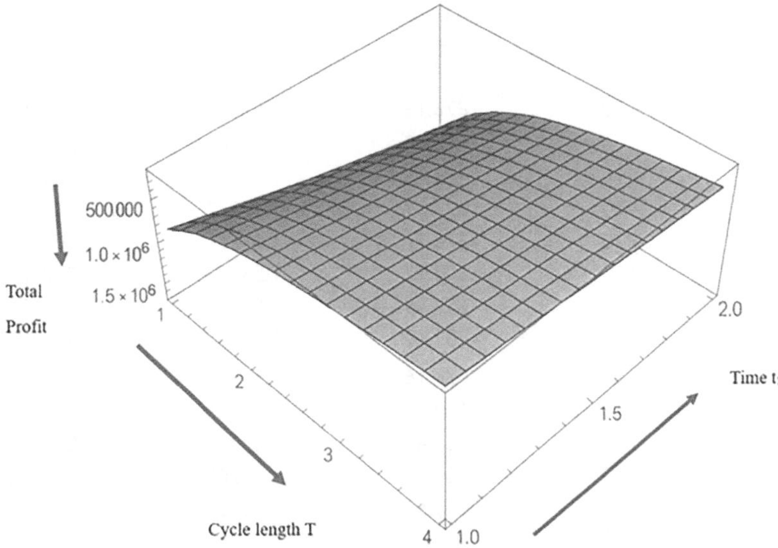

Figure 5.2 Concavity graph between cycle length T, t_1, and total profit.

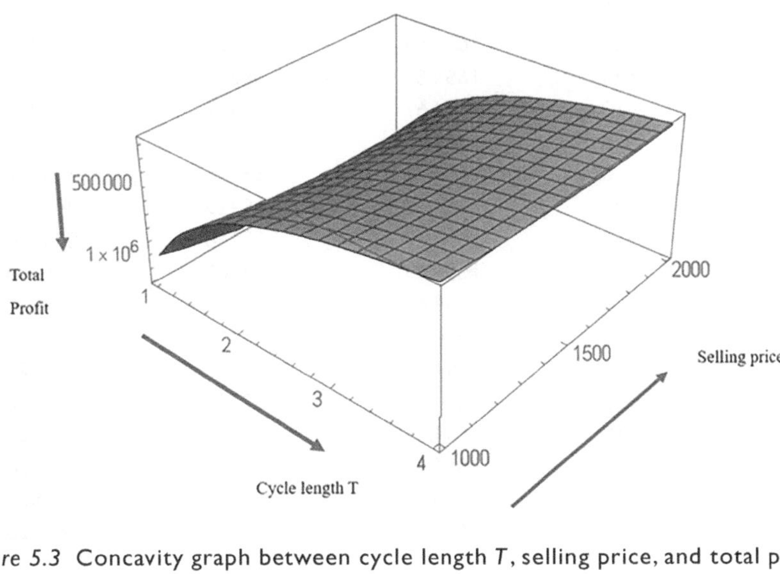

Figure 5.3 Concavity graph between cycle length T, selling price, and total profit.

5.7 SENSITIVITY ANALYSIS

The behavior of different parameters is shown in Table 5.2.

Table 5.2 Sensitivity analysis of various parameters

Parameter	*% value*	*Selling price p*	*Cycle length T*	*Total profit TP*
a	+20	1987.45	1.5904	1213660
	+10	1828.18	0.674726	124348×10^{11}
	0	1654.49	1.53768	844520
	−10	1488.03	1.50969	684979
	−20	1321.57	1.48047	542125
B	+20	1376.71	1.53808	705640
	+10	1502.97	1.53788	768767
	0	1654.49	1.53768	844520
	−10	1852.62	1.54993	937298
	−20	2071.16	1.53728	1.05×10^{6}
h	+20	1652.41	1.53742	846717
	+10	1653.45	1.53753	845618
	0	1654.49	1.53768	844520
	−10	1655.53	1.5379	843422
	−20	1656.58	1.53819	842325
h'	+20	1654.35	1.53766	844666
	+10	1654.42	1.53767	844593
	0	1654.49	1.53768	844520
	−10	1654.56	1.5377	844447
	−20	1654.63	1.53771	844373
θ	+20	1654.84	1.48247	843267
	+10	1654.67	1.50991	843890
	0	1654.49	1.53768	844520
	−10	1654.31	1.56581	845157
	−20	1654.12	1.59429	845801
G	+20	1654.49	1.53768	844320
	+10	1654.49	1.53768	844420
	0	1654.49	1.53768	844520
	−10	1654.49	1.53768	844620
	−20	1654.49	1.53768	844720
γ	+20	1654.49	1.53769	844520
	+10	1654.49	1.53769	844520
	0	1654.49	1.53768	844520
	−10	1654.49	1.53768	844520
	−20	1654.49	1.53768	844520

(*Continued*)

Table 5.2 (Continued)

Parameter	*% value*	*Selling price p*	*Cycle length T*	*Total profit TP*
γ'	+20	1654.49	1.53769	844520
	+10	1654.49	1.53768	844520
	0	1654.49	1.53768	844520
	−10	1654.49	1.53768	844520
	−20	1654.49	1.53768	844520
ψ	+20	1654.49	1.53775	844517
	+10	1654.49	1.53772	844518
	0	1654.49	1.53768	844520
	−10	1654.49	1.53765	844521
	−20	1654.49	1.53762	844522
ω	+20	1654.49	1.53741	844531
	+10	1654.49	1.53755	844525
	0	1654.49	1.53768	844520
	−10	1654.49	1.53782	844514
	−20	1654.49	1.53795	844509
ω'	+20	1654.49	1.53741	844527
	+10	1654.49	1.53755	844523
	0	1654.49	1.53768	844520
	−10	1654.49	1.53782	844516
	−20	1654.49	1.53795	844513

5.8 OBSERVATIONS

1. As the demand scale parameter "*a*" increases, the total profit, cycle length, and selling price all rise accordingly. However, an anomaly is observed when the scale parameter reaches 1100.
2. As the price elasticity coefficient "*b*" increases, the total profit, selling price, and cycle length all decrease.
3. As holding costs increase, total profit, selling price, and cycle length decrease.
4. As energy costs in the holding process increase, total profit increases, while selling price and cycle length decrease.
5. As the deterioration rate increases, the total profit decreases.
6. As green technology investment increases, the total profit decreases while cycle length and selling price remain the same, but sustainability will increase.
7. As the product development cost cycle length increases, but the profit and price of sale remain the same.
8. On an increase in the energy cost to develop a sustainable product system, the cycle length increases, but the profit and price of sale remain the same.

9. On an increase in the material cost per unit, total profit decreases and cycle length increases, but selling price remains the same.
10. On an increase in the tool/die cost, the cycle length decreases, but total profit increases.
11. On an increase in the renewable energy cost during production, the cycle length decreases, but total profit increases.

5.9 MANAGERIAL INSIGHTS

1. Managers seek to reduce the environmental effects of the manufacturing system while striving to achieve environmentally friendly production.
2. Managers should try to avoid the scale parameter 1100 for more benefit, that is manager can get more profit.
3. Managers should keep the price elasticity coefficient to a minimum for getting more profit.
4. Managers should try to keep the deterioration rate to a minimum.
5. Managers should invest in green technology for sustainable development.

5.10 CONCLUSION

In this chapter, we formulate an inventory model for liquid products like medicine and oils. In this chapter, demand for products is considered to be dependent on selling price. Green technology investment is considered to reduce the effect of carbon emissions. To avoid the shortage, the flexible production rate is taken into account. We use renewable energy for sustainable development in the storage and production processes. Numerical analysis is carried out using software Mathematica 12.0. From the sensitivity analysis, we observe that when the demand scale parameter is greater than 1100, the profit increases. Also, we obtain that as the investment in green technology increases, the total profit will decrease, but sustainable development will increase. This study can be further extended by using different carbon emission policies, stochastic demand rate, and under a fuzzy environment.

ACKNOWLEDGMENT

The first author would like to thank CCS University Meerut, for providing financial help in the form of URGC with letter no. fmo. Dev/1043 dated 29/06/22.

REFERENCES

Datta, T. K. (2017). Effect of green technology investment on a production inventory system with carbon tax. *Advances in Operations Research*, 2017, Article ID 4834839, 12.

Hasan, M. R., Roy, T. C., Daryanto, Y., & Wee, H. M. (2021). Optimizing inventory level and technology investment under a carbon tax, cap-and-trade and strict carbon limit regulations. *Sustainable Production and Consumption*, 25, 604–621.

Jaggi, C. K., Priyamvada, & Kamna, K. M. (2022). Sustainable production system with preservation strategy and renewable energy under different carbon tax policies. *International Journal of Modelling and Simulation*, 43, 1–10.

Kök, A. G., Shang, K., & Yücel, Ş. (2018). Impact of electricity pricing policies on renewable energy investments and carbon emissions. *Management Science*, 64(1), 131–148.

Rahman, M. S., Das, S., Manna, A. K., Shaikh, A. A., Bhunia, A. K.,Cárdenas-Barrón, L. E., & Céspedes-Mota, A. (2021). A mathematical model of the production inventory problem for mixing liquid considering preservation Facility. *Mathematics*, 9(24), 3166.

Rajabi, H., Mosleh, M. H., Mandal, P., Lea-Langton, A., & Sedighi, M. (2020). Emissions of volatile organic compounds from crude oil processing–Global emission inventory and environmental release. *Science of the Total Environment*, 727, 138654.

Sarkar, B., Omair, M., & Choi, S. B. (2018a). A multi-objective optimization of energy, economic, and carbon emission in a production model under sustainable supply chain management. *Applied Sciences*, 8(10), 1744.

Sarkar, M., & Chung, B. D. (2021). Effect of renewable energy to reduce carbon emissions under a flexible production system: A step toward sustainability. *Energies*, 14(1), 215.

Sarkar, M., Sarkar, B., & Iqbal, M. W. (2018b). Effect of energy and failure rate in a multi-item smart production system. *Energies*, 11(11), 2958.

Sepehri, A., Mishra, U., Tseng, M. L., & Sarkar, B. (2021). Joint pricing and inventory model for deteriorating items with maximum lifetime and controllable carbon emissions under permissible delay in payments. *Mathematics*, 9(5), 470.

Sharma, S., Singh, S. R., & Kumar, M. (2021). A reverse logistics inventory model with multiple production and remanufacturing batches under fuzzy environment. *RAIRO - Operations Research*, 55(2), 571–588.

Singh, S. R., Yadav, D., Sarkar, B., & Sarkar, M. (2021). Impact of energy and carbon emission of a supply chain management with two-level trade-credit policy. *Energies*, 14(6), 1569.

Vitale, I., Dondo, R. G., González, M., & Cóccola, M. E. (2022). Modelling and optimization of material flows in the wood pellet supply chain. *Applied Energy*, 313, 118776.

Chapter 6

Linear and weakly nonlinear stability analysis of thermal convection in Jeffrey ferromagnetic fluid

Fourier series truncated approximation

Kavita, Veena Sharma, and Dinesh Bhatia

6.1 INTRODUCTION

It is well known fact that a ferrofluid is a highly magnetized fluid under the influence of a magnetic field. At least three elements are required to manufacture a ferrofluid: carrier liquid, colloidal magnetic particles, and stabilizer (surfactant). These are colloidal single-domain ferromagnetic particles suspended in a nonmagnetic carrier liquid with a size of 10 nm. When the particle size is less than 1–2 nm, the magnetic properties of permanently magnetized nanoparticles diminish and colloidal motion increases as the particle size decreases. Ferrofluids are widely used in various magnetic fluid-based scientific equipment, such as sensors, densimeters, actuating machines, pressure transducers, and accelerometers, such as energy converters and electromechanical converters (Odenbach, 2009).

Finlayson (1970) investigated the convective instability of a ferromagnetic fluid confined to a layer heated from below. He demonstrated the theory of thermomechanical interaction among ferromagnetic fluids for both rigid horizontal and shear-free boundaries using the linear stability method. Rosensweig (1985) provided a detailed introduction to ferrohydrodynamics. Aggarwal and Prakash (2009) investigated the influence of suspended particles and rotation on the onset of ferrofluids using the linear stability theory.

In recent years, ferrofluids have received significant research attention owing to their diverse applications across several scientific and technological domains, including medical sciences, instrumentation, vacuum technology, oscillation damping, loudspeaker cooling, and high-speed silent printing devices. Notably, ferrofluids play a crucial role in magnetically targeted drug delivery systems, enabling the controlled transport of therapeutic agents—such as radionuclides, cancer-specific antibodies, and genes—to localized regions of the human body. These techniques are particularly effective in cancer therapy, detoxification processes, and the in vivo monitoring of brain chemical activity (Goodwin et al., 1999; Pulfer and Gallo, 2020).

Ferrofluids exhibit additional important properties due to the magnetic field, such as adhesion properties, magneto-viscosity, and non-Newtonian

 DOI: 10.1201/9781003521389-6

properties. Rosenblat (1986) investigated thermal convection in a layer of viscoelastic fluid heated from below and generalized the classical Lorenz system of Newtonian–Bénard problem. Joseph and Preziosi (1989) proposed elastico-viscous fluids, which exhibit linear viscoelastic features, yield stress, shear-thinning characteristics, and high shear viscosity. These fluids are known as Jeffrey fluids and exhibit non-Newtonian properties. Shahnazian et al. (2019) investigated the viscoelastic properties of ferromagnetic fluids, focusing on the formation of chain-like clusters of magnetic nanoparticles using nonspherical cobalt nanoparticles. Borin and Odenbach (2009) examined the rheological properties of two kinds of ferromagnetic layers and provided an estimated impact of magneto-viscosity by taking into account the temperature dependence of magnetization in a frozen sample.

The linear and nonlinear methods for unstable processes persist clouded in controversies. However, after almost a century of efforts in nonlinear theory, the investigation and calculation of such nonlinear stability problems became possible by considering nonlinear components in the equations of motion and energy and by using a model dynamo capable of generating nonlinear finite amplitude. In weakly nonlinear analysis, the original system exhibits an approximate or asymptotic solution close to the bifurcation point. In terms of a small parameter, the solution is enlarged around a basic state. Both time and spatial coordinates may influence the nonlinear growth of a disturbance. To capture the modulation, s multiple scales (of time or spatial dimensions) are introduced to the system, along with an amplitude expansion of the solution. Many investigators proposed various approaches to solve the nonlinear differential equations defined for regular fluids in the monographs (Craik, 1968; Squire, 1933; Zhou, 1982). Stuart (1960) introduced a two-dimensional theory that reduces the exponentially periodic growth to accommodate nonlinear influences in the early 1960s. Later, Benney and Chow (1989) proposed a more effective nonlinear theory known as the first harmonic theory of mean flow. Using linear and nonlinear theories, Qin and Kaloni (1994) investigated the buoyancy–surface tension impact in a magnetic fluid heated from below.

A decade later, Macek and Strumik (2010) succeeded in generalizing the Lorenz model for inclined horizontal magnetic fluid, taking into account a new Kelvin force induced by the magnetic field. Laroze et al. (2010) investigated theoretically and numerically the amplitude equation for stationary convection considering the non-Newtonian base fluid. Rahman and Suslov (2015) studied the linear stability analysis of magnetoconvection of ferrofluids in the presence of a uniform external magnetic field, and nonlinear stability was analyzed due to the influence of magnetic field in ferrofluids. The dynamics and complexity in the generalized Lorenz system for nonlinear thermal convection in a horizontal magnetic viscous fluid were examined by Macek (2018). Siddheshwar et al. (2019) examined the initiation of thermal convection in a rotating ferromagnetic and dielectric fluid using the Lorenz model, and they found that the critical Rayleigh number of the linear stability theory was compared with that obtained by the weakly nonlinear stability method. The influence of temperature-dependent viscosity on thermal convection in a ferrofluid layer was investigated by Dhiman and Sharma (2021), whereas linear and weakly nonlinear stability analysis of oscillatory convection in the presence of Coriolis force was analyzed by Dhiman and Sood (2022). In their research, they have found that magnetic number and magnetization have a destabilizing impact on ferro-convection. Dev and Suthar (2022) found that the Rayleigh number for two different sets of realistic boundary conditions in the presence of internal heat on the onset of ferrofluids and analyzed nonlinear stability using the Chebyshev Tau method, where they observed that the magnetic parameter advanced the onset of penetrative convection.

Motivated by the above-mentioned facts, this chapter aims to analyze the linear and weakly nonlinear stability of thermal convection in rheological ferrofluids and is primarily motivated by the work of Siddheshwar et al. (2019). The rheological behavior of the base fluid is described by the Jeffrey model. In this study, the generalized Lorenz model is used for nonlinear equations and is evaluated numerically using the software MATHEMATICA (version-12). A comparison is made in detail for the nonlinear stability boundaries and linear instability. A coincidence is observed between nonlinear and linear results in the absence of buoyancy magnetization for base fluids, which implies the exclusion of the occurrence of subcritical stability.

6.2 MATHEMATICAL FORMULATION

An infinitely extending, electrically nonconducting, incompressible horizontal layer of viscoelastic Jeffrey ferromagnetic fluid heated from below between horizontal parallel planes; $z = 0$, $z = d$ is considered. The physical system is pervaded by a homogeneous magnetic field, $\mathbf{H}_f\left(0,0,H\right)$ and gravitational force, $\mathbf{g}\left(0,0,-g\right)$ along the z-axis.

For mathematical simplification, all the involved physical variables are assumed to be independent of the y-coordinate, thereby yielding two-dimensional ferro-convection. Since the horizontal planes are infinitely extended, the dynamics of the base fluid is described by the bulk of the fluid, which remains unaffected by the lateral wall $\left(x\right)$.

The appropriate mathematical equations of continuity, momentum, energy, and density for viscoelastic Jeffrey ferromagnetic fluid (Finlayson, 1970; Siddheshwar et al., 2019) using the Boussinesq approximation are as follows:

$$\nabla \cdot \boldsymbol{u} = 0, \tag{6.1}$$

$$\rho_0\left[\frac{\partial \boldsymbol{u}}{\partial t} + \left(\boldsymbol{u}\cdot\nabla\right)\boldsymbol{u}\right] = -\nabla p_f + \rho\mathbf{g} + \mu_0\left(\mathbf{M}_f\cdot\nabla\right)\boldsymbol{H}_f + \frac{\mu}{1+\lambda}\nabla^2\boldsymbol{u}, \tag{6.2}$$

$$\frac{\partial T}{\partial t}+(\boldsymbol{u}\cdot\nabla)T=\kappa\nabla^2 T, \tag{6.3}$$

$$\rho_f=\rho_0\left[1-\alpha_f\left(T-T_0\right)\right], \tag{6.4}$$

where $\boldsymbol{u}$, t, ρ_0, μ_0, μ, λ, T, κ, α_f, T_0, $\boldsymbol{H}_f$, and $\boldsymbol{M}_f$ denote, respectively, the velocity vector, the time, the static density, the magnetic permeability, the fluid viscosity, the Jeffrey parameter (defined as the ratio of relaxation and retardation time), the temperature, the thermal diffusivity, the thermal expansion coefficient, the reference temperature, the magnetic field, and the magnetization.

In equation (6.2), small-scale convective motion is only considered, therefore, $\left|(\boldsymbol{u}\cdot\nabla)\boldsymbol{u}\right| << 1$. Thus equation (6.2) is modified as

$$\rho_0\frac{\partial \boldsymbol{u}}{\partial t}=-\nabla p_f+\rho \mathrm{g}+\mu_0\left(\boldsymbol{M}_f\cdot\nabla\right)\boldsymbol{H}_f+\frac{\mu}{1+\lambda}\nabla^2\boldsymbol{u}, \tag{6.5}$$

Maxwell's equations are

$$\nabla\times\boldsymbol{H}_f=0,\quad \nabla\cdot\boldsymbol{B}=0. \tag{6.6}$$

Further, the magnetic induction, $\boldsymbol{B}$, is related to magnetic field, $\boldsymbol{H}_f$ and magnetization, $\boldsymbol{M}_f$, and is defined by

$$\boldsymbol{B}=\mu_0\left(\boldsymbol{H}_f+\boldsymbol{M}_f\right). \tag{6.7}$$

Magnetic equation of state is given by (Finlayson, 1970)

$$M_f\left(T,H_f\right)=M_0-K_1\left(T-T_0\right)+\chi_1\left(H_f-H_0\right), \tag{6.8}$$

where $K_1=-\left(\frac{\partial M}{\partial T}\right)_{H_0,T_0}$ and $\chi_1=-\left(\frac{\partial M}{\partial H}\right)_{H_0,T_0}$ represent the coefficients of pyromagnetic and magnetic susceptibility, respectively, $M_0=M_f\left(T_0,H_0\right)$ denotes reference magnetization.

Also, magnetic field, $\boldsymbol{H}_f$, in terms of magnetic potential, ϕ, can be written as

$$H_f=\nabla\phi. \tag{6.9}$$

Since the boundary conditions are taken to be stress-free. Therefore, the appropriate boundary conditions are as follows:

$$w=\frac{\partial^2 w}{\partial z^2}=0, T=T_1 \text{ and } T=T_0 \text{ at } z=0 \text{ and } z=d. \tag{6.10}$$

It is expected that both the tangential component of the magnetic field and the normal component of magnetic induction are continuous. Therefore,

$$D\phi + \frac{a\phi}{1+\chi_1} - T = 0 \text{ at } z = 0,$$

$$D\phi - \frac{a\phi}{1+\chi_1} - T = 0 \text{ at } z = d. \tag{6.11}$$

In equation (6.11), as $\chi_1 \to \infty$, we get $D\phi \to 0$ at $z = 0$, $z = d$, which is an isothermal boundary condition for temperature (Finlayson, 1970).

The basic state to assume is quiescent and steady. Therefore,

$$\boldsymbol{u} = \boldsymbol{u}_b(0,0,0), p_f = p_b(z), T_b(z) = T_0 - \beta(z-d), \rho_f = \rho_b(z),$$

$$\boldsymbol{M}_f = \boldsymbol{M}_b(z), \boldsymbol{H}_f = \boldsymbol{H}_b(z). \tag{6.12}$$

where $\boldsymbol{M}_b = \left\{ M_0 + \frac{K_1\beta(z-d)}{1+\chi_1} \right\} \hat{\boldsymbol{k}},$

$$\boldsymbol{H}_b = \left\{ H_0 - \frac{K_1\beta(z-d)}{1+\chi_1} \right\} \hat{\boldsymbol{k}}, \tag{6.13}$$

where $\hat{\boldsymbol{k}} = (0,0,1)$.

To examine the stability of the system, extremely small perturbations are superimposed on the basic state of the physical variables solutions as follows:

$$\boldsymbol{u} = \boldsymbol{u}_b + \boldsymbol{u}', p_f = p_b + p', T = T_b + T', \rho_f = \rho_b + \rho',$$
$$\boldsymbol{M}_f = \boldsymbol{M}_b + \boldsymbol{M}', \boldsymbol{H}_f = \boldsymbol{H}_b + \boldsymbol{H}', \tag{6.14}$$

the prime symbols represent the usual meaning.

Using perturbations (6.14) in Maxwell's equations (6.7) and after simplification, we get

$$\frac{\partial}{\partial x}(H' + M') = \frac{\partial H'}{\partial x}\left(\frac{H_0 + M_0}{H_0}\right), \tag{6.15}$$

$$\frac{\partial}{\partial z}(H' + M') = \frac{\partial H'}{\partial z}(1+\chi_1) - K_1 T'. \tag{6.16}$$

By substituting the perturbations given in equation (6.14) into equations (6.1), (6.3), and (6.5), we get the relevant nonlinear perturbed equations of the ferrofluid as follows:

$$\nabla \cdot \boldsymbol{u}' = 0, \tag{6.17}$$

$$\rho_0 \frac{\partial \boldsymbol{u}'}{\partial t} = -\nabla P_1 + \rho_0 \alpha T' g \hat{\boldsymbol{k}} + \frac{\mu}{1+\lambda} \nabla^2 \boldsymbol{u}' + \mu_0$$

$$\left[\nabla\left(\phi'_z M_b\right) - K_1 T' \nabla \phi'_z\right] - \mu_0 \left[K_1 \beta \phi'_z - \frac{K_1{}^2 \beta T'}{1+\chi_1}\right] \hat{\boldsymbol{k}}, \tag{6.18}$$

where the modified pressure term is $P_1 = P_f - \frac{\mu_0 M_0}{2H_0} \phi_x'^2 - \frac{\mu_0 \chi_1}{2} \phi_z'^2$.

$$\frac{\partial T'}{\partial t} + \left(\boldsymbol{u}' \cdot \nabla\right) T' = \kappa \nabla^2 T' + \beta w'. \tag{6.19}$$

Nondimensional variables are introduced

$$\left(x^*, y^*, z^*\right) = \frac{(x, y, z)}{d}, \boldsymbol{u}^* = \frac{d}{\kappa} \boldsymbol{u}', T^* = \frac{1}{\beta d} T', \phi^* = \frac{1+\chi_1}{\beta d^2 K_1} \phi', P^* = \frac{d^2}{\mu \kappa} P_1,$$

$$H^* = \frac{H_f}{H_0}, M^* = \frac{M_f}{M_0}, \tag{6.20}$$

Using equations (6.15)–(6.19) and using linear stability theory in the dimensionless linear form (omitting * for the sake of simplicity), we obtain

$$\nabla \cdot \boldsymbol{u} = 0, \tag{6.21}$$

$$\frac{1}{p_r} \frac{\partial \boldsymbol{u}}{\partial t} = -\nabla P_1 + R_f \left(1 + M_1\right) T \hat{k} - R_f M_1 \phi_z \hat{k} + \frac{1}{1+\lambda} \nabla^2 \boldsymbol{u} - R_f M_1 T \nabla \phi_z, \tag{6.22}$$

$$\frac{\partial T}{\partial t} + \left(\boldsymbol{u}.\nabla\right) T = \nabla^2 T + w, \tag{6.23}$$

$$M_3 \nabla_1^2 \phi + \phi_{zz} - T_z = 0. \tag{6.24}$$

Here,

$$p_r = \frac{\mu}{\rho_0 \kappa}, R_f = \frac{\rho_0 \alpha \Delta T d^3 g}{\mu \kappa}, M_1 = \frac{\mu_0 K_1{}^2 \Delta T}{\rho_0 d \left(1+\chi_1\right) g}, M_3 = \frac{H_0 + M_0}{H_0 \left(1+\chi_1\right)}$$

are the Prandtl number, the thermal Rayleigh number of the ferrofluid, the ratio of the magnetic to buoyancy forces, and non-buoyancy magnetization (measuring the departure of linearity in the magnetic equation of state), respectively, which are nondimensional parameters.

Eliminating the pressure, P_1, from the horizontal components of equation (6.22) by applying suitable operators and using equation (6.21), one gets

$$\frac{1}{p_r}\frac{\partial}{\partial t}\nabla^2 w = \frac{1}{1+\lambda}\nabla^4 w + R_f\left(1+M_1\right)\nabla_1^2 T - R_f M_1 \nabla_1^2 \phi_z$$
$$-R_f M_1 \frac{\partial}{\partial x}\frac{\partial\left(T,\phi_z\right)}{\partial\left(x,z\right)}, \tag{6.25}$$

where $\dfrac{\partial\left(T,\phi_z\right)}{\partial\left(x,z\right)} = \dfrac{\partial T}{\partial x}\dfrac{\partial \phi_z}{\partial z} - \dfrac{\partial T}{\partial z}\dfrac{\partial \phi_z}{\partial x}$ is the Jacobian term of T and ϕ_z.

The boundary conditions (6.10) using equation (6.20) transform to

$$w = D^2 w = 0, T = D\phi = 0 \text{ at } z = 0 \text{ and } z = 1. \tag{6.26}$$

6.2.1 Linear stability analysis

For linear stability analysis, the perturbations in physical variables are analyzed into normal modes and are taken in the following form:

$$\left[w', T', \phi'\right] = \left[W\left(z\right), \Theta\left(z\right), \Phi\left(z\right)\right] e^{\left\{i\left(k_x x + k_y y\right) + \varpi t\right\}}, \tag{6.27}$$

Here, $\varpi = \omega_r + \iota\omega_i$ is the growth rate of disturbance, k_x and k_y are the respective horizontal wavenumbers, along x- and y-directions, $a\left[= \left(k_x^2 + k_y^2\right)^{\frac{1}{2}}\right]$ is the resultant perturbed wave number.

Using expression (6.27), equations (6.23)–(6.25) take the form:

$$\frac{\varpi}{p_r}\left(D^2 - a^2\right)W = \frac{1}{1+\lambda}\left(D^2 - a^2\right)^2 W - a^2 R_f\left(1+M_1\right)\Theta$$
$$+a^2 R_f M_1 D\Phi, \tag{6.28}$$

$$\varpi\Theta = \left(D^2 - a^2\right)\Theta + W, \tag{6.29}$$

$$D^2\Phi - a^2 M_3 \Phi - D\Theta = 0. \tag{6.30}$$

The set of coupled equations (6.28)–(6.30) with boundary conditions (6.26) constitutes an eigenvalue problem whose solutions ought to be found. Using the single-term Galerkin technique, the exact solutions of equations (6.28)–(6.30) with boundary conditions (6.26) characterizing the lowest mode are chosen as

$$W = A_1 \sin \pi z,\ \Theta = B_1 \sin \pi z,\ \Phi = -C_1 \frac{1}{\pi} \sin \pi z, \tag{6.31}$$

where A_1, B_1, and C_1 are constants.

Using these solutions in equations (6.28)–(6.30) and integrating each equation by parts individually over $z = 0$ to $z = 1$ and utilizing boundary conditions (6.26), one gets the matrix form as follows:

$$\begin{bmatrix} \left(\frac{\varpi}{p_r} + \frac{1}{1+\lambda}(\pi^2 + a^2)\right)(\pi^2 + a^2) & -a^2 R_f (1 + M_1) & a^2 R_f M_1 \\ 0 & -1 & \left(M_3 \frac{a^2}{\pi^2} + 1\right) \\ -1 & \varpi + (\pi^2 + a^2) & 0 \end{bmatrix} \begin{bmatrix} A_1 \\ B_1 \\ C_1 \end{bmatrix} = 0. \tag{6.32}$$

For equation (6.32) to admit a nontrivial solution, the determinant of the associated coefficient matrix must vanish, leading to the determination of the thermal Rayleigh number R_f of Jeffrey ferrofluid as

$$R_f = \frac{(a^2 + \pi^2)(a^2 M_3 + \pi^2)(a^2 + \pi^2 + \varpi)\left[(a^2 + \pi^2) p_r + \varpi(1 + \lambda)\right]}{a^2\left[a^2(1 + M_1) M_3 + \pi^2\right] p_r (1 + \lambda)} \tag{6.33}$$

which on putting $\varpi = i\omega_i \neq 0$ gives after simplifications

$$R_f = \Delta_1 + i\omega_i \Delta_2. \tag{6.34}$$

Here,

$$\Delta_1 = \frac{(a^2 + \pi^2)(a^2 M_3 + \pi^2)\left[(a^2 + \pi^2)^2 p_r - \omega_i^2 (1 + \lambda)\right]}{a^2\left[a^2(1 + M_1) M_3 + \pi^2\right] p_r (1 + \lambda)}, \tag{6.35}$$

and

$$\Delta_2 = \frac{(a^2 + \pi^2)^2 (a^2 M_3 + \pi^2)[1 + p_r + \lambda]\omega_i}{a^2\left[a^2(1 + M_1) M_3 + \pi^2\right] p_r (1 + \lambda)}. \tag{6.36}$$

R_f, being a physical parameter, must take real values. Equation (6.34) implies either the occurrence of stationary modes $\omega_i = 0$ or that of oscillatory modes $\omega_i \neq 0$.

6.2.2 Stationary convection

The occurrence of stationary cellular convection is described by taking $\omega_i = 0$ in equation (6.35), which yields the thermal Rayleigh number of stationary modes as follows:

$$R_f^s = \frac{\left(a^2+\pi^2\right)^3\left(a^2M_3+\pi^2\right)}{a^2\left[a^2\left(1+M_1\right)M_3+\pi^2\right]\left(1+\lambda\right)}. \tag{6.37}$$

It is noteworthy from equation (6.37) that R_f^s depends on non-buoyancy magnetization parameter M_3, ratio of the magnetic to buoyancy forces M_1, and Jeffrey parameter λ.

Now special cases arise:

Case (I): For regular ferrofluid, that is, $\lambda = 0$, equation (6.37) reduces to

$$R_f^s = \frac{\left(a^2+\pi^2\right)^3\left(a^2M_3+\pi^2\right)}{a^2\left[a^2\left(1+M_1\right)M_3+\pi^2\right]}, \tag{6.38}$$

which is exactly the same as that of Finlayson (1970), which gives the interaction of the buoyancy and magnetic modes of instability.

Case (II): For regular Newtonian fluid, i.e., $M_3 = 0$ and $M_1 = 0$, equation (6.38) further condenses to

$$R_f^s = \frac{\left(a^2+\pi^2\right)^3}{a^2},$$

which is in coincidence with the earlier result for the classical Bénard convection (Chandrashekhar, 1961).

To examine the effect of the ratio of the magnetic to buoyancy forces M_1, non-buoyancy magnetization parameter M_3, and Jeffrey parameter λ, on the stability of stationary convection of viscoelastic Jeffrey ferrofluids, we examine the behavior of $\frac{\partial R_f^s}{\partial M_1}$, $\frac{\partial R_f^s}{\partial M_3}$, and $\frac{\partial R_f^s}{\partial \lambda}$ analytically.

Equation (6.37) yields

$$\frac{\partial R_f^s}{\partial M_1} = -\frac{M_3\left(a^2+\pi^2\right)^3\left(a^2M_3+\pi^2\right)}{\left[a^2\left(1+M_1\right)M_3+\pi^2\right]^2\left(1+\lambda\right)}. \tag{6.39}$$

$$\frac{\partial R_f^s}{\partial M_3} = -\frac{M_1\pi^2\left(a^2+\pi^2\right)^3}{\left[a^2\left(1+M_1\right)M_3+\pi^2\right]^2\left(1+\lambda\right)}. \tag{6.40}$$

$$\frac{\partial R_f^s}{\partial \lambda} = -\frac{\left(a^2+\pi^2\right)^3\left(a^2M_3+\pi^2\right)}{a^2\left[a^2\left(1+M_1\right)M_3+\pi^2\right]\left(1+\lambda\right)^2}. \tag{6.41}$$

It is assessed from equations (6.39)–(6.41) that $\frac{\partial R_f^s}{\partial M_1}$, $\frac{\partial R_f^s}{\partial M_3}$, and $\frac{\partial R_f^s}{\partial \lambda}$ are negative for all wave numbers. Therefore, the parameters M_1, M_3, and λ always show a destabilizing effect on a system.

6.2.3 Oscillatory convection

The thermal Rayleigh number of oscillatory modes is given by comparing the real and imaginary components of equation (6.34) as

$$R_f^{osc} = \frac{\left(a^2+\pi^2\right)\left(a^2M_3+\pi^2\right)\left[\left(a^2+\pi^2\right)^2 p_r - \omega_i^2\left(1+\lambda\right)\right]}{a^2\left[a^2\left(1+M_1\right)M_3+\pi^2\right]p_r\left(1+\lambda\right)}, \tag{6.42}$$

and the frequency of oscillatory modes is described by the following expression

$$\frac{\left(a^2+\pi^2\right)^2\left(a^2M_3+\pi^2\right)\left[1+p_r+\lambda\right]}{a^2\left[a^2\left(1+M_1\right)M_3+\pi^2\right]p_r\left(1+\lambda\right)}\omega_i = 0. \tag{6.43}$$

Therefore, equation (6.43) implies either $\frac{\left(a^2+\pi^2\right)^2\left(a^2M_3+\pi^2\right)\left[1+p_r+\lambda\right]}{a^2\left[a^2\left(1+M_1\right)M_3+\pi^2\right]p_r\left(1+\lambda\right)} = 0$ or $\omega_i = 0$. But $\frac{\left(a^2+\pi^2\right)^2\left(a^2M_3+\pi^2\right)\left[1+p_r+\lambda\right]}{a^2\left[a^2\left(1+M_1\right)M_3+\pi^2\right]p_r\left(1+\lambda\right)} \neq 0$. Therefore, $\omega_i = 0$ which implies that oscillatory modes do not exist. Consequently, oscillatory convection is ruled out, and the principle of exchange of stabilities is valid for linear stability analysis in ferromagnetic fluids.

6.2.4 Weakly nonlinear analysis using the Lorenz approximation model

This weakly nonlinear analysis is carried out by assuming the truncated Fourier series solutions for the equations (6.23)–(6.25) with the help of the Lorenz model in the following form (Laroze et al., 2010):

$$w(x,z,t) = aA_2(t)\cos(ax)\sin(\pi z),$$

$$T(x,z,t) = B_2(t)\cos(ax)\sin(\pi z) + B_3(t)\sin(2\pi z),$$

$$\phi(x,z,t) = C_2(t)\cos(ax)\cos(\pi z) - 2\pi C_3(t)\sin(2\pi z), \tag{6.44}$$

where $A_2(t), B_2(t), B_3(t), C_2(t)$, and $C_3(t)$ are finite amplitudes.

Using these solutions from equation (6.44) in equation (6.24), one gets the values of $C_2(t)$ and $C_3(t)$ in terms of $B_2(t)$ and $B_3(t)$, respectively, as

$$C_2(t) = -\frac{\pi}{J_M^2} B_2(t) \text{ and } C_3(t) = -\frac{1}{2\pi} B_3(t), \tag{6.45}$$

where $J_M^2 = (\pi^2 + M_3 a^2)$.

Again, substituting the solutions given by equation (6.44), equations (6.22) and (6.25), and using the orthogonality conditions on the system of equations, we get

$$\frac{d}{dt} A_2(t) = -\frac{J^2}{1+\lambda} p_r A_2(t) + a p_r R_f B_2(t)\left[\zeta_1 - \zeta_2 B_3(t)\right], \tag{6.46}$$

$$\frac{d}{dt} B_2(t) = A_2(t)\left[\pi a B_3(t) + a\right] - J^2 B_2(t), \tag{6.47}$$

$$\frac{d}{dt} B_3(t) = -\frac{\pi a}{2} A_2(t) B_2(t) - 4\pi^2 B_3(t), \tag{6.48}$$

where

$$J^2 = (\pi^2 + a^2),\ \zeta_1 = \frac{\left(J_M^2 + M_1 M_3 a^2\right)}{J^2 J_M^2} \text{ and } \zeta_2 = \frac{\left(\pi M_1 M_3 a^2\right)}{J^2 J_M^2}. \tag{6.49}$$

6.2.5 Steady finite-amplitude convection

Since steady finite convection amplitudes are independent of time, eliminating $B_3(t)$ and $B_2(t)$ from equations (6.47) and (6.48), using equation (6.46), one gets the governing equation to establish a steady-state two-dimensional finite amplitude after simplifications as

$$\alpha_1 \left(A_2^2\right)^2 + \alpha_2 A_2^2 + \alpha_3 = 0. \tag{6.50}$$

Here, $\alpha_1 = \pi J^2 a^4$, $\alpha_2 = 8\pi a^4 R_f \zeta_1 (1+\lambda) + 8a^4 R_f \zeta_2 (1+\lambda) - 16\pi a^2 J^4$, and $\alpha_3 = 64\pi a^2 J^2 R_f \zeta_1 (1+\lambda) - 64\pi J^6$.

The minima of steady, finite-amplitude thermal Rayleigh number R_f on making a discriminant of equation (6.48) vanish, which gives

$$R_{af} = \frac{4M_1M_3\left(a^2M_3 + \pi^2\right)\left(a^2 + \pi^2\right)^3}{\left[a^2\left(M_3 + 2M_1M_3\right) + \pi^2\right]^2 (1+\lambda)}. \tag{6.51}$$

6.3 NUMERICAL RESULTS AND DISCUSSION

The non-oscillatory thermal Rayleigh number, R_f for linear and steady for finite-amplitude Rayleigh number, R_{af} for nonlinear stability analysis encapsulated in equations (6.37) and (6.51). The thermal Rayleigh numbers with varying wave number are calculated using the software MATHEMATICA-12 by varying one of the nondimensional parameters keeping others fixed at permissible values, as considered by Dev and Suthar (2022); Siddheshwar et al. (2019); Mahajan and Parashar (2020), and many more, that is, $M_1 = 1, \lambda = 0.6$, and $M_3 = 10$. These computations have been represented graphically. In each of the figures, NL and L represent linear stability and weakly nonlinear stability, respectively.

Figures 6.1 and 6.2 plot the thermal Rayleigh number, R_f variations versus wavenumber, a for the linear stationary convection and for finite-amplitude perturbations (nonlinear steady convection), respectively, for various values of the ratio of the magnetic to buoyancy forces, $M_1 = 1,2,3$ and non-buoyancy magnetization parameter, $M_3 = 5,15,30$. It is assessed from the figures that the thermal Rayleigh number shows a fall with a rise in the parameter M_1 and M_3, implying thereby the destabilizing influence on a system.

Figure 6.3 illustrates the effect of R_f with a for distinct values of the viscoelastic parameter, $\lambda = 0.2,0.4,0.6$. The thermal Rayleigh number decreases with a rise in the viscoelastic parameter, thereby depicting the destabilizing impact of the Jeffrey parameter on the system.

The validation of the present results with viscous ferrofluids, that is, $\lambda = 0$, with linear stability analysis of Finlayson (1970) and weakly nonlinear

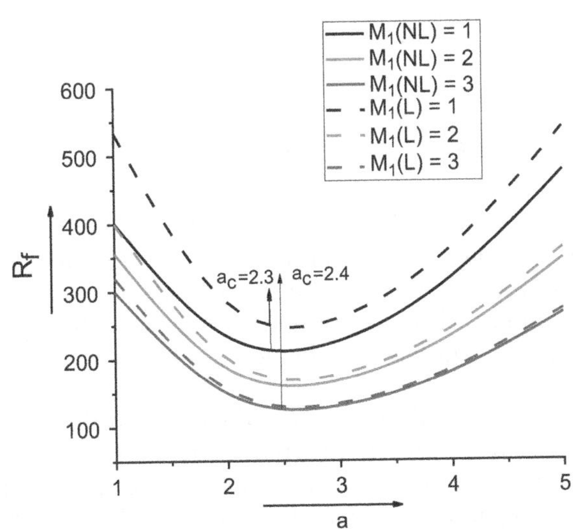

Figure 6.1 Variation of R_f versus a as a function of M_1.

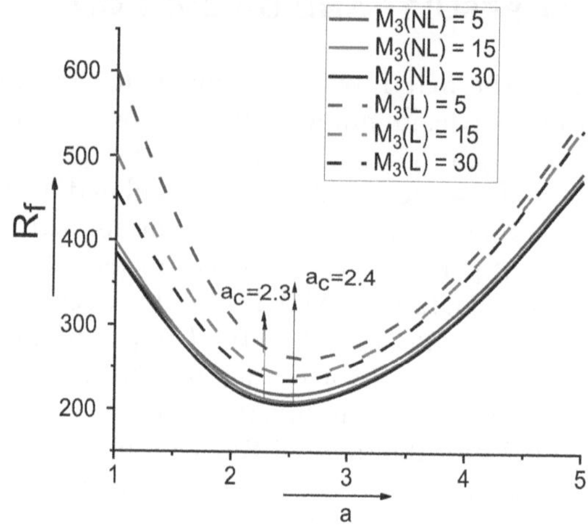

Figure 6.2 Variation of R_f versus a as a function of M_3.

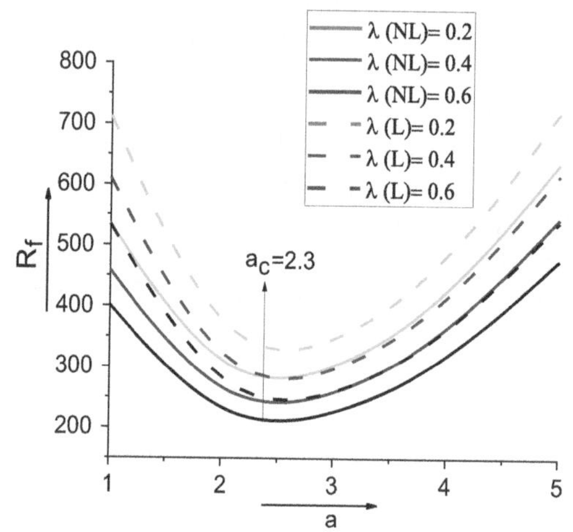

Figure 6.3 Variation of R_f versus a as a function of λ.

stability analysis of Siddheshwar et al. (2019), is carried out in Figure 6.4. For a smaller buoyancy magnetization parameter, M_1, the linear and finite-amplitude analysis predicts a large difference in the subcritical motion of thermal Rayleigh number, whereas approximately the same Rayleigh number and the same critical wave number are found for larger buoyancy magnetization parameter M_1.

Figures 6.5 and 6.6 plot the variation of critical thermal Rayleigh number, R_{fc} against viscoelastic parameter, $\lambda = 0.2, 0.4, 0.6$ for a fixed critical wave number $a_c = 2.3$. It is depicted from the figures that critical thermal Rayleigh decreases with an increase in parameters M_1 and M_3, respectively. Thus, M_1 and M_3 tend

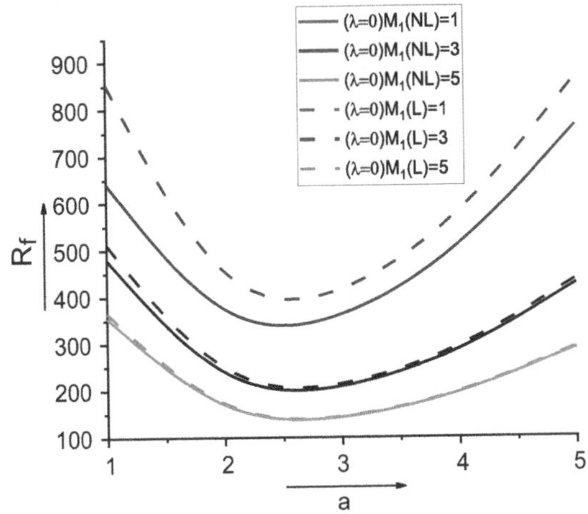

Figure 6.4 Variation of R_f versus a as a function of M_1.

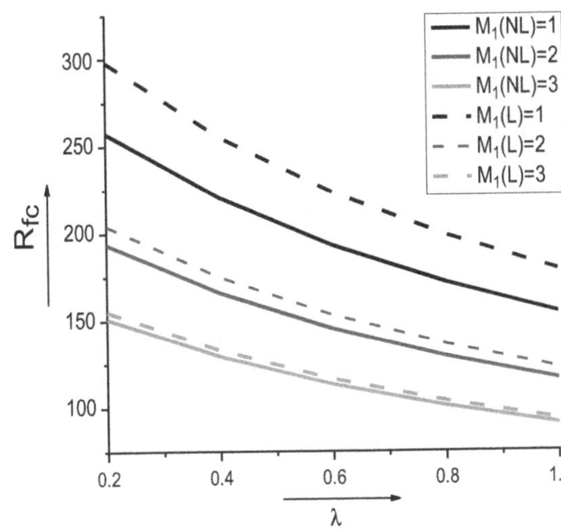

Figure 6.5 Variation of R_{fc} versus parameter λ as a function of M_1.

to shrink stability region under stationary modes, thereby advancing the initiation of stationary convection for both linear and nonlinear stabilities.

The critical thermal Rayleigh number R_{fc} against distinct values of the parameters M_1 and M_3, respectively, is shown in Figures 6.7 and 6.8. Very interesting results are observed from both the figures that critical Rayleigh number decreases with an increase in the values of M_1 for linear stability, depicting thereby its destabilizing effect, that is, advancing the start of stationary convection. However, a reverse trend occurs for variation with M_3, that is, the critical thermal Rayleigh number first increases for $M_3 < 1$, then

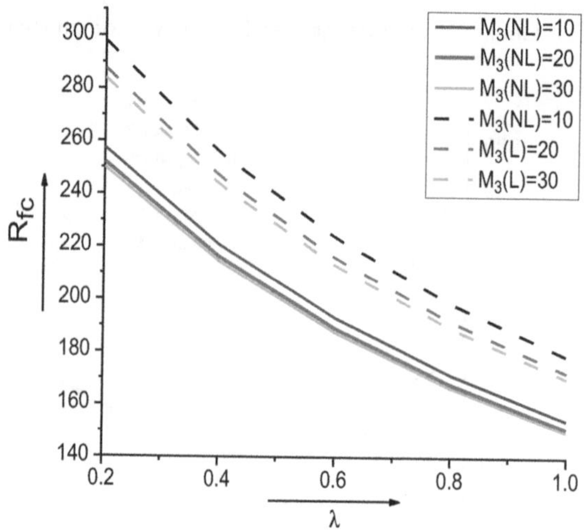

Figure 6.6 Variation of R_{fc} versus parameter λ as a function of M_3.

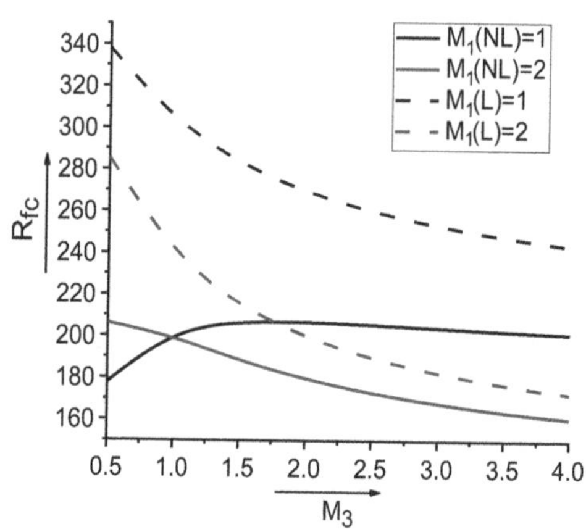

Figure 6.7 Variation of R_{fc} versus parameter M_3 as a function of M_1.

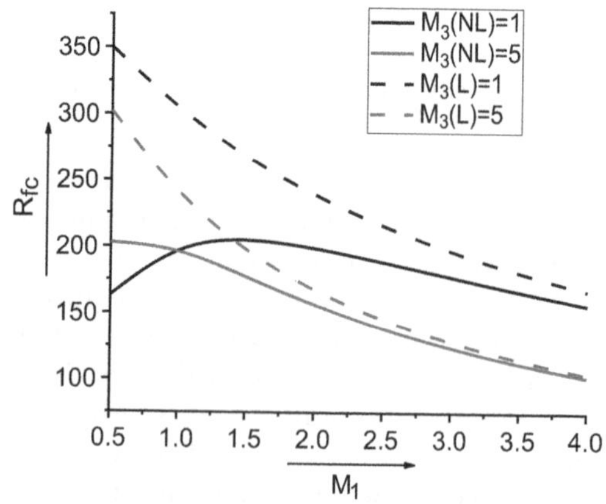

Figure 6.8 Variation of R_{fc} versus parameter M_1 as a function of M_3.

starts to decrease for $M_1 = 1$. The critical Rayleigh number decreases with an increment in M_3 for $M_1 = 2$. Consequently, this was expected from Figure 6.4. These results are similar to those of Siddheshwar et al. (2019).

6.4 CONCLUSION

The onset of the ferrofluid layer is investigated for the Jeffrey model using the linear and weakly nonlinear stability analysis. The three-dimensional problem of linear partial differential equations is solved by the Galerkin technique, and the two-dimensional problem of nonlinear equations is solved using truncated Fourier series demonstration. The following conclusions are made from this chapter:

1. The occurrence of oscillatory motion is ruled out.
2. The critical thermal Rayleigh number of stationary ferromagnetic convection decreases with an increase in both the ratio of the magnetic to buoyancy forces, M_1 and non-buoyancy magnetization parameter, M_3. Hence, buoyancy forces and non-buoyancy magnetization advance the start of stationary convection in both M_1 and M_3.
3. Jeffrey parameter λ destabilizes a system and prepones the initiation of stationary convection.
4. The nonlinear stability thermal Rayleigh number for non-ferrofluid at the globe is exactly the same as that of the linear instability thermal Rayleigh number. The low value of the magnetic parameter causes this coincidence to disappear right away.

REFERENCES

Aggarwal, A. K., and Prakash, K. 2009. "Effect of suspended particles and rotation on thermal instability of ferrofluids." *International Journal of Applied Mechanics and Engineering* 14 (1): 55–66.

Benney, D. J., and Chow, K. 1989. "A mean flow first harmonic theory for hydrodynamic instabilities." *Studies in Applied Mathematics* 80 (1): 37–73.

Borin, D. Y., and Odenbach, S. 2009. "Magnetic measurements on frozen ferrofluids as a method for estimating the magnetoviscous effect." *Journal of Physics: Condensed Matter* 21 (24): 246002.

Chandrashekhar, S. 1961. *Hydrodynamics and hydromagnetic stability*. Oxford University Press.

Craik, A. D. 1968. "A note on the static stability of an elastico-viscous fluid." *Journal of Fluid Mechanics* 33 (1): 33–38.

Dev, K., and Suthar, O. P. 2022. "Nonlinear stability analysis of penetrative convection in ferrofluids via internal heating." *Journal of Magnetism and Magnetic Materials* 564 (2): 170064.

Dhiman, J. S., and Sharma, N. 2021. "Effect of temperature dependent viscosity on thermal convection in ferrofluid layer." *Journal of Theoretical and Applied Mechanics Sofia* 51 (1): 3–21.

Dhiman, J. S., and Sood, S. 2022. "Linear and weakly non-linear stability analysis of oscillatory convection in rotating ferrofluid layer." *Applied Mathematics and Computation* 430: 127239.

Finlayson, B.A. 1970. "Convective instability of ferromagnetic fluids." *Journal of Fluid Mechanics* 40 (4): 753–767.

Goodwin, P. J., Ennis, M., Pritchard, K. I., Trudeau, M., and Hood, N. 1999. "Risk of menopause during the first year after breast cancer diagnosis." *Journal of Clinical Oncology* 17 (8): 2365–2365.

Joseph, D. D., and Preziosi, L. 1989. "Heat waves." *Reviews of Modern Physics* 61 (1): 41–73.

Laroze, D., Martinez-Mardones, J., and Pérez, L. M. 2010. "Amplitude equation for stationary convection in a viscoelastic magnetic fluid." *International Journal of Bifurcation and Chaos* 20 (2): 235–242.

Macek, W. M. 2018. "Nonlinear dynamics and complexity in the generalized Lorenz system." *Nonlinear Dynamics.* 94 (4): 2957–2968.

Macek, W. M., and Strumik, M. 2010. "Model for hydromagnetic convection in a magnetized fluid." *Physical Review E* 82 (2): 027301.

Mahajan, A., and Parashar, H. 2020. "Linear and weakly nonlinear stability analysis on a rotating anisotropic ferrofluid layer." *Physics of Fluids* 32 (2): 024101.

Odenbach, S. Ed. 2009. *Colloidal magnetic fluids: basics, development and application of ferrofluids.* Springer, 763.

Pulfer, S. K. and Gallo, J. M. 2020. "Targeting tumors using magnetic drug delivery." In *Biomedical chemistry: applying chemical principles to the understanding and treatment of disease.* John Wiley and Sons, Inc., 211–225.

Qin, Y., and Kaloni, P. N. 1994. "Nonlinear stability problem of a ferromagnetic fluid with surface tension effect." *European Journal of Mechanics - B/Fluids* 13 (3): 305–321.

Rahman, H., and Suslov, S. A. 2015. "Thermomagnetic convection in a layer of ferrofluid placed in a uniform oblique external magnetic field." *Journal of Fluid Mechanics* 764: 316–348.

Rosenblat, S. 1986. "Thermal convection in a viscoelastic liquid." *Journal of Non-Newtonian Fluid Mechanics* 21 (2): 201–223.

Rosensweig, R. E. 1985. *Ferrohydrodynamics.* Cambridge University Press.

Shahnazian, H., Gräf, D., Borin, D. Y., and Odenbach, S. 2019. "Rheology of a ferrofluid based on nanodisc cobalt particles." *Journal of Physics D: Applied Physics* 1 (12): 1–11.

Siddheshwar, P. G., Suthar, O. P., and Chinnaswamy, K. 2019. "Finite-amplitude ferro-convection and electro-convection in a rotating fluid." *SN Applied Sciences* 1 (12): 1–11.

Squire, H. B. 1933. "On the stability for three-dimensional disturbances of viscous fluid flow between parallel walls." *Proceedings of the Royal Society of London. Series A, Containing Papers of a Mathematical and Physical Character* 142 (847): 621–628.

Stuart, J. T. 1960. "On the non-linear mechanics of wave disturbances in stable and unstable parallel flows Part 1. The basic behaviour in plane Poiseuille flow." *Journal of Fluid Mechanics* 9 (3): 353–370.

Zhou, H. 1982. "On the nonlinear theory of stability of plane Poiseuille flow in the subcritical range." *Proceedings of the Royal Society of London. A. Mathematical and Physical Sciences.* 381 (1781): 407–418.

Chapter 7

A new approach for solving generalized function in fractional calculus rate

Raghvendra Singh, Rajendra Singh, and P N Pathak

7.1 INTRODUCTION

The function $F = F_q(a,t)$ provides a simple response to the fundamental straight fractional-order difference, as well as a comprehensive understanding of the scenario and its associated initial value difficulty. This chapter studies the Laplace transforms of related functions, including the R-function and the G-function. These functions are helpful to computations in both analytical and fractional calculus. The R-function differs from the F-function because it includes all F-derivatives and integrals. In addition to returning its own value at the qth level of differences, the R-function additionally returns its own answer at the qth order.

Generalized functions are objects in mathematics that are extensions of the concept of functions. There are several recognized theories, such as the theory of distributions, which is one example. Generalized functions are especially useful for translating discontinuous functions into continuous functions and for characterizing discrete phenomena such as point charges. In addition, they are important for understanding discrete physical phenomena like point charges. A set of all regular sequences with highly well-behaved functions constitutes a structure analogous to a maintenance sequence. A distribution is also referred to as a "generalized function" or a "ideal function" in some circles. Generally speaking, a generalized function is a generalization of the idea of a function, as its name implies.

Generalized measures are known as systematic linear functionals over an universe of infinitely differentiable functions, such that the derivatives of all continuous functions are also generalized functions. They are related to long-term linear exchange-correlation functional over a space of indefinitely differentiable functions, where all continuous functions possess generalized derivatives. The delta function is the generalized function that is encountered the most frequently.

It would be highly advantageous to design a general function that returns itself as the result of fractionally variable integration (of any order). Utilizing

DOI: 10.1201/9781003521389-7

this method would ease the study of the second differential significantly. To accomplish this objective, the following suggestions are given:

$$R_{q,v}[a,c,t]=\sum_{n=0}^{\infty}\frac{(a)^{n}(t-c)^{(n+1)q-1-v}}{\Gamma(n+1)(q-v)}$$

Regarding partial differential equations, we are interested primarily in the equations of fractional derivatives in the region $t > c = 0$. R will be real for $t - c$, unless the exponent is zero

$$R_{q,v}[a,t-c]=\sum_{n=0}^{\infty}\frac{(a)^{n}(t-c)^{(n+1)q-1-v}}{\Gamma(n+1)(q-v)}$$

where $(n+1)(q-1-v)$ is really an integer. Additionally, the smaller, more compact version is preferable, especially when $c = zero$:

The transform operator is used for the R-function:

$$L\{R_{q,v}[a,c,t]\}=L\sum_{n=0}^{\infty}\frac{(a)^{n}(t-c)^{(n+1)q-1-v}}{\Gamma(n+1)(q-v)}$$

$$L\{R_{q,v}[a,c,t]\}=\sum_{n=0}^{\infty}(a)^{n}L\frac{(t-c)^{(n+1)q-1-v}}{\Gamma(n+1)(q-v)}$$

Consider the first situation where $c = 0$, then we obtain

$$L\{R_{q,v}[a,0,t]\}=\sum_{n=0}^{\infty}(a)^{n}L\frac{(t)^{(n+1)q-1-v}}{\Gamma(n+1)(q-v)}$$

According to Erdelyi et al. (1954),

$$L\{t^{v}\}=\Gamma(v+1)s^{-v-1},\quad \mathrm{Re}(v)>-1,\mathrm{Re}(s)>0.$$

Then, the value of the equation becomes

$$L\{R_{q,v}[a,0,t]\}=\sum_{n=0}^{\infty}(a)^{n}\frac{1}{S^{(n+1)}(q-v)}\qquad \mathrm{Re}\{(n+1)q-v\}>-1,$$

$$\mathrm{Re}(s)>0.$$

$$L\left\{R_{q,v}\left[a,0,t\right]\right\} = \frac{1}{S^{-v}} \sum_{n=0}^{\infty} \frac{(a)^n}{S^{(n+1)}q} \qquad \operatorname{Re}\left\{(n+1)q - v\right\} > 0, \operatorname{Re}(s) > 0.$$

This may be expressed mathematically as a vector space that merges when $a / Sq = 1$. It may be demonstrated with the long division method:

$$L\{R_{q,v}(a,0,t\} = \frac{S^v}{S^q - a} \qquad \operatorname{Re}(q - v) > 0, \operatorname{Re}(s) > 0.$$

The Laplace transform's shifting theorem for $c \neq 0$ is now

$$L\left\{f(t-b)u(t-b)\right\} = e^{-bs} L\left\{f(t)\right\} \; b \geq 0.$$

The one-step functional $u(t-b)$ allows $(t-b)$ to sum to zero whenever t and b are given. This mathematics and its consequences provide, assuming $R_{q,v}\left[a,c,t\right] = 0 \, \text{for} \, tc$:

$$L\{R_{q,v}(a,c,t\} = \frac{e^{-cs} S^v}{S^q - a} \qquad c \geq 0, \quad \operatorname{Re}(n+1)(q-v) > 0, \operatorname{Re}(s) > 0.$$

7.2 FRACTIONAL CALCULUS

Over the years, various mathematicians have devised numerous formulations that correspond to the concept of a non-integer graded integral or derivative, each using their own notation and technique. The Riemann–Liouville declaration is a popular variant of the Riemann–Liouville theorem in fractional calculus. It is noteworthy that, although having distinct definitions, the Riemann–Liouville and Lacroix definitions of a nonlinear fractional yield the same result.

Fraction calculus is a field of mathematics derived from proportional-derivative arithmetic techniques, similar to how fractional coefficients evolved from the ideas of integer exponents. Consider the exponent's physical importance. In the context of applied mathematics, fractional calculus is the theory and application of arbitrary order integration and differentiation. Despite being incorrect, the term fractional has lasted due to its frequent use. It is an area of mathematics that examines the many options for determining the power and impact of the differentiated operator D on real or complex numbers.

$$Df(x) = \frac{d}{dx} f(x),$$

Moreover, J is the integration operator

$$Jf(x) = \int_0^x f(s)ds,$$

As well as constructing a calculus for such operators that is a generalization of the classical calculus. In this context, empowerment refers to the repeated trials of a systems are operating D to a function f, or the act of repeatedly constructing D with itself, as in

$$D^n(f) = \underbrace{D \circ D \circ D \circ \cdots \circ D}_{n}(f) = \underbrace{D(D(D(\cdots D}_{n}(f)\cdots))).$$

7.2.1 Fractional calculus functionalities

This section provides fractional differential equations functions that have proven beneficial for addressing fractional calculus issues, including partial differential equation and fractional differential equation solutions.

7.2.2 Agarwal's purpose

In (1953), the Mittag–Leffler functional is given by Agarwal:

$$E_{\alpha,\beta}(t) = \sum_{m=0}^{\infty} \frac{t^{\left(m+\frac{\beta-1}{\alpha}\right)}}{\Gamma((\alpha m+\beta))}$$

This feature is especially essential in fractional system theory due to its transfer function, as stated by Agarwal:

$$L\left\{E_{\alpha,\beta}\left[t^{\alpha}\right]\right\} = \frac{S^{\alpha-\beta}}{S^{\alpha}-1}$$

The function is the small section derivatives of the F-function, as described by Hartley and Lorenzo (1998) and Robotnov (1969), as $(a = 1)$ the parameter. It will be addressed in the future.

7.2.3 Erdelyi's function

Erdelyi et al. (1954) analyzed the implementation of the Mittag–Leffler operator as shown below:

$$E_{\alpha,\beta}(t) = \sum_{m=0}^{\infty} \frac{t^{(m)}}{\Gamma((\alpha m+\beta))} \qquad \alpha,\beta > 0$$

In this instance, the integer factors of t are considered. Given that function, the transform may be expressed as follows:

$$L\{E_{\alpha,\beta}(t)\} = \sum_{m=0}^{\infty} \frac{\Gamma(m+1)}{\Gamma(\alpha m+\beta)S^{m+1}} \qquad \alpha,\beta > 0.$$

This procedure will not be investigated further due to the difficulty of generalizing it.

7.2.4 The Robotnov–Hartley function

The following components were created to provide the sole solution to the basic linear fractional derivative (Hartley and Lorenzo 1998):

$$F_q\{a,t\} = t^{q-1}\sum_{n=0}^{\infty} \frac{(-a)^n t^{nq}}{\Gamma(nq+1)} q > 0.$$

Robotnov (1969, 1980). A similar study was conducted on the concept of hereditary integrals and their application to solid design. The robustness and elegance of this function's transfer function, namely its inverse, are two of its most distinguishing qualities:

$$L\{F_q\{a,t\}\} = \frac{1}{S^q - a} \quad ,q > 0$$

7.2.5 Miller and Ross' function

A new function is presented as an alternative to the function provided to Miller and Ross (1993, pp. 80 and 309–351) for the solution of the finite difference starting value problem. As stated above, it is the vth integral of the exponent:

$$E_t(v,a) = \frac{d^{-v}}{dt^{-v}} e^{at} = t^v e^{at} \gamma^*(v,at) = t^v \sum_{k=0}^{\infty} \frac{(at)^k}{\Gamma(v+k+1)}$$

(v,at) denotes the incompletion of the function. The transfer function represented by equation may be stated as

$$L\{E_t(v,a)\} = \frac{S^{-v}}{S-a} \qquad \mathrm{Re}(v) > 1.$$

Miller and Ross demonstrate, therefore, that

$$L\left\{\sum_{j=1}^{q} a^{j-1}E_t\left(jv-1,a^q\right)\right\} = \frac{1}{S^v - a}, q = 1,2,3.....,v = \frac{1}{q} = 1,\frac{1}{2},\frac{1}{3},...$$

It is a special instance of Robotnov and Hartley's F-function.

7.3 R-FUNCTION

The R-function and its counterpart, the F-function, are mathematical functions that are fundamental to fractional mathematics because they allow fractions to be stated in fractional form. The solutions to the foundational fraction linear difference equation may be written in terms of the purpose described in this section, as demonstrated. These computations act as enhancements to the exponential growth employed in the partial differential equation's solution.

The F-function is represented by

$$E_q(a,t) \equiv \sum_{n=0}^{\infty} \frac{a^n t^{(n+1)q-1}}{\Gamma\left[(n+1)q\right]}$$

In addition to its generalization, the R-function is also represented by

$$R_{q,v}(a,c,t) \equiv \sum_{n=0}^{\infty} \frac{a^n t^{(n+1)q-1-v}}{\Gamma\left[(n+1)q-v\right]}$$

These concepts are of crucial significance in fractional calculus and depicts several characteristics of the R-function. Using $v = 0$ and $a = 1$ to illustrate the effect of q alterations using these parameters, Figure 7.1 illustrates the influence of q modifications when $v = 0$.

The function's exponential nature is easily obvious (see $q = 1$). The effect is a result that begins when $t = c = 0$:

$$L\left\{R_{q,v}(a,c,t)\right\} = \frac{S^v}{S^q - a}$$

It is also noted that

$$R_{1,0}(a,0,t) = R_{1,0}(1,0,at) = \sum_{n=0}^{\infty} \frac{(at)^n}{\Gamma(n+1)} = e^{at}$$

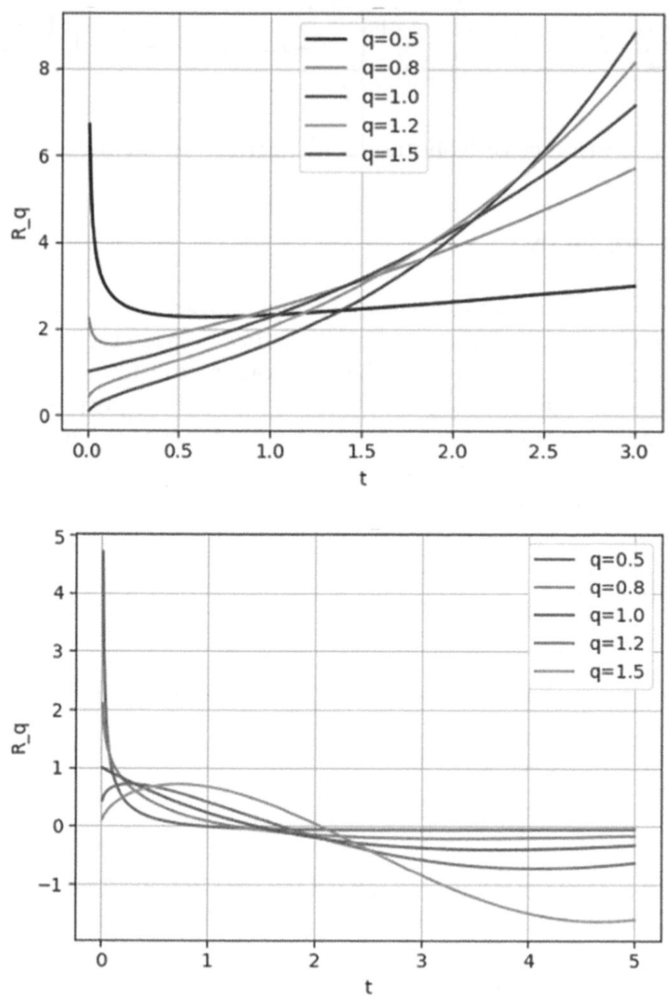

Figure 7.1 (a) Effect of q on $R_{q,0}(1,0,t), v=0.0, a=1.0$ and (b) effect of q on $R_{q,0}(-1,0,t), v=0.0, a=-1.0$.

When it comes to the analysis that follows, these relationships will be very valuable! This is special, because in general

$$R_{q,v}(a,0,t) \neq R_{q,v}(1,0,at)$$

The following useful relationship, on the other hand, is shown:

$$R_{q,v}(1,0,at) = \sum_{n=0}^{\infty} \frac{(at)^{(n+1)q-1-v}}{\Gamma[(n+1)q-v]} = a^{q-1-v} \sum_{n=0}^{\infty} \frac{(a^q)^n t^{(n+1)q-1-v}}{\Gamma[(n+1)q-v]}$$

Therefore,

$$R_{q,v}(1,0,at) = a^{q-1-v} R_{q,v}\left(a^q,0,t\right)$$

Alternatively, this may be written as

$$R_{q,v}(a,0,t) = a^{\frac{(v+1-q)}{q}} R_{q,v}\left(1,0,a^{\frac{1}{q}}t\right)$$

7.4 CONCLUSION

This study shows that the fractional calculus R-function can be derived. Unlike the F-function, the R-function contains all the integrals and derivatives related to the F-function. Eigen showed that the R-function reproduces itself through qth – order fractional differentiation and integration. Important examples of the R-function include the activation curve, sine and cosine functions, and the corresponding hyperbolic sine and hyperbolic cosine functions. For a dynamic thermistor, the R-function enables direct transformation of harmonic-domain results into time-periodic solutions without requiring additional equipment.

BIBLIOGRAPHY

Agarwal, R.P.. On the Mittag-Leffler function. *Proceedings of the National Institute of Sciences of India*, 19, 41–47, 1953.

Erdelyi, A., Magnus, W., Oberhetinger, F., and Tricomi, F.G. (Eds.), *Tables of Integral Transforms*, vol. 1, McGraw-Hill Book Co. (1954), LCC Number 54-6214.

Hartley, T.T. and Lorenzo, C.F., A solution to the fundamental linear fractional order differential equation, *NASA/TP*—1998-208963 (December 1998).

Miller, K.S. and Ross, B., *An Introduction to the Fractional Calculus and Fractional Differential Equations*. John Wiley & Sons, Inc. (1993).

Robotnov, Y.N., *Tables of a Fractional Exponential Function of Negative Parameters and Its Integral (in Russian)*. Nauka (1969).

Robotnov, Y.N., *Elements of Hereditary Solid Mechanics (in English)*. MIR Publishers (1980).

Chapter 8

Analysis of Quality of Services (QoS) for wireless sensor network

Vinam Tomar, Vinti Gupta, and Abhinav Goel

8.1 INTRODUCTION

The wireless sensor system (Baghyalakshmi et al. 2011) consists of different nodes that can communicate with nature by detecting or controlling physical parameters. These nodes need to work together. Nodes are connected through a wireless connection, forming a wireless sensor network (WSN), with sensors arranged at each node to collect data and can work together to meet the combined goal. Sensor nodes work in threatening situations, for example, in fighters and maintenance zones. Due to its operational nature, unusual WSN attacks and many new types of attacks are introduced.

8.1.1 Sensor nodes

A sensor node is a network issue to that detects and then connects the data. Depending on the algorithm used, the sensor starts moving the operation and/or questions based on the concerning node transaction. Depending on the needs of the system, the node can perform several calculations. After the calculation, it may transfer the data to the next data. In the sphere of the sensor node, the sensor can function as a source or sink/actuator.

8.1.2 Gateways

Scientists and system administrators can use the gateway to connect motors to personal computers (PCs), personal digital assistants (PDAs), the Internet, and existing networks and protocols. In other words, gateway acts as a proxy for the sensor network on the Internet. It can function in activate, inactive, or distributed hybrid mode and allows the active door sensor node to send data to a gateway server.

8.1.3 Task managers

The work manager has linked the media to the Gateway Task Manager (such as Internet or satellite links), such as data services, client data browsing, and processing. These work organizers can be seen as an information

DOI: 10.1201/9781003521389-8

recovery and processing platform. All information from sensor nodes is stored in the work manager for data analysis (raw, filtered, and processed). The user can find/analyze this local or remote information using any display interface (i.e., PDA and computer).

8.2 CHARACTERISTICS OF WSN

To compare the traditional WSN configuration system, such as MANET (mobile Ad hoc networks) and simple cellular systems, the following features and limitations are observed:

- Battery-Enabled (Dependent) Sensor Nodes
  The sensor nodes are basically battery-powered and kept in an environment where the battery change and charging are difficult.
- Self-Organized
  The placement of nodes has been configured randomly and is automatically configured as a communication network.
- Malicious Sensor Nodes
  Suspected Nodes in Sensor nodes are posted due to physical defects or harsh environments.
- Storage Constraints, Delivered Energy, and Computation
  Sensor Nodes are capable of a limited library, accountability, and ability. The sensor node's energy, calculation, and storage capabilities are very limited.
- Redundancy in data
  To manage applications for maximum sensors, the sensor hub is firmly located in the area and cooperates to achieve general sensing work.

8.3 LITERATURE SURVEY

Xu et al. (2018) and Cirstea et al. (2012) studied on a two-neuron system, based on digital analysis and hardware experiences, in a non-autonomous Hopfield Neural Network (HNN), exhibiting dual dynamical behavior. The stability of the DC competitive viewpoint is examined and discovered in a parameter region of interest that is centered on a non-stable point. Animated maps are used to uncover stimulation-linked dynamic behavior data instead of buffering diagrams and two-dimensional parameters, which can have a dual appealing growth attitude with a variety of two-linked attachment characteristics. In addition, breadboard experiences are investigated, which validate the judicial simulations (Rebentrost et al. 2018). Quantum computing has significant developmental capabilities in terms of speed and the ability to utilize widely used machine learning techniques. The author has shown that the network can be stored as a large network in the number of commuted bits. After introducing a classic technique to run the Hopfield network, logical algorithms are applied in the dimensions of the data to achieve a quantitative sculpture of that quantity. They also offer our procedure for identifying our genetic layout (Bao et al. 2017). With three neurons, an easy-to-use connection topology, HNN, is offered as a type of NHN-based nonlinear system. An integrated adjustment parameter is a dynamic behavior with a variety of adjusted parameters using weight loss and traditional dynamic analysis methods as parameters and the behavior of multiple attraction behavior under various states. The starting values are validated. The results reveal that the HNN-based system exhibits point of view, length, odd behavior, doubling, and transit traffic. This technique, in particular, exhibits various incentives to promote many motivations, such as a single scroll encourages a couple of exceptional sensations, as well as a pair of anticipation, a pair of two dimensions predicted desirable, and so on. Experimental measurement results are very satisfactory and satisfied with pace contexts (Yang et al. 2017). An HNN-based system's hardware circuit may be built using widely available electrical components, and physical experiences can capture a wide range of affiliate sentiments. Contemporary circuits have been presented based on some precious monuments. However, implementing some easy neural network circuits is still not included, at least. This chapter is part of the construction of the novel and memorable Hopfield network. A neural network circuit is a type of circuit. While considering the PP spice process, a memorable bridge is employed to complete the circuit handling function, which is zero, positive, and negative without any switch or invoicing. Completing the flow of the circuit simplifies the network structure and slows down the process of converting current and voltage signals. In addition, in binary and color pictures, the association memory is shown based on the recommended memorable network (Patel and Mistry 2017). A WSN is a tiny wireless sensor node network. Sensors are used to detect physical or environmental factors. Therefore, such attacks are needed to protect the network (Cabeza et al. 2016). The most artificial neural network model now uses vital information for training. Information may be inconvenient for application errors in industrial systems, such as data processing and data retrieval, and may be less accurate or unverified due to various factors. In this regard, the HNN diagnostic system has been proposed to address this problem. With successful performance, the analysis was tested using the development and application of actor diagnostic practices in Industrial Control System Indicators (DAMADICS) and achieved a successful performance (Duan et al. 2016). In this chapter, we systematically designed the current association of neural networks, including the role of the biological world and recent monuments in the traditional HNN. In particular, the originally fully connected Hopfield network is slow to consider the small global impact, based on the priority of removing priority connections, that is, weight slip. Built-in network association has a low connection on balanced performance. In addition, a hardware plan for small-world Hopfield networks is analyzed by commercially adaptive makers using synaptic-based circuits. Most of them have similar capabilities to the network; each node has the same giving output power, storage, energy, and capabilities. There are two methods of the WSN node Advanced Node (high energy) and the common nodes. The node connector is not used until the recommended network work is in existence. In addition, when the battery is loaded in unfavorable surroundings, it is difficult to fill the battery node. Hu et al. (2015) show associate memory based on a memorable Hopfield network. Rejecting the goals of different pattern monuments can be stored in the Hopfield network, and prestored designs could be effectively accessed directly or through Associated States Intermediate states, which have a companion's behavior. Wang et. al. (2015a) Single Association Memory and Multi Association Memories can be found with memorable Hopfield networks (Zhang et al. 2015). Practically, HNN are used as a sampling to communicate how to process every stage of processed information (Gupta and Seth, 2022). To demonstrate the reliability of implementation information, an analysis of the stability of these systems is performed. The important point is, these common results can help build the construction of laptop works, which are used by the analysis of multiple-layer stability of the low-HNN (Agarwal et al. 2022). Consequently, a combination of considerable conditions is obtained to ensure stability. In addition, the common conditions of continuous or timely external inputs depend on the case that these networks meet in case the stability conditions can be easily used to achieve complete and minor synchronization (Yadav 2021). Finally, two judicial examples have been presented to show the effectiveness of our ideological results.

8.4 PROPOSED ALGORITHM

```
while()
Create a node network set in a circle
Select the source and destination sensor node
if(sensor node detects collision)
then
Identify the wireless network to identify the node pattern
Start transmission from the source node again
else
Transmit the data to all nodes
End if
End while
```

8.5 RESULTS AND DISCUSSION

This compares the performance of each classifier with the transfer rate, the end of the end, and the throughput. We introduced these standards to better understand the results.

Packet delivery ratio: Representation of the absolute number of messages distributed by each spectator hub, divided by the full number of messages created by allocation centers of the opportunity is included. It can be set by very well-compatible equations. Figure 8.1 represents the Packet Drop Scenario on a round basis. The Packet Drop ratio analysis for this manuscript is demonstrated in Figure 8.2.

$$\text{PDR} = \left(\left(\text{total packets} - \text{loss}\right) / \text{total packets}\right) * 100$$

End2End Delay—The system has a specified time for a packet when the parcel takes the source and achieves the goal. Figure 8.3 depicts the end-to-end delay analysis for this manuscript.

Throughput—A fixed number of invoice times is the number of packages passing through a channel. This performance measure indicates that data is effectively sent from the source node to the destination node and may be enhanced by increasing node density. The throughput analysis in this chapter is depicted in Figure 8.4. In response to a given question, the sample size generated by the network is equal

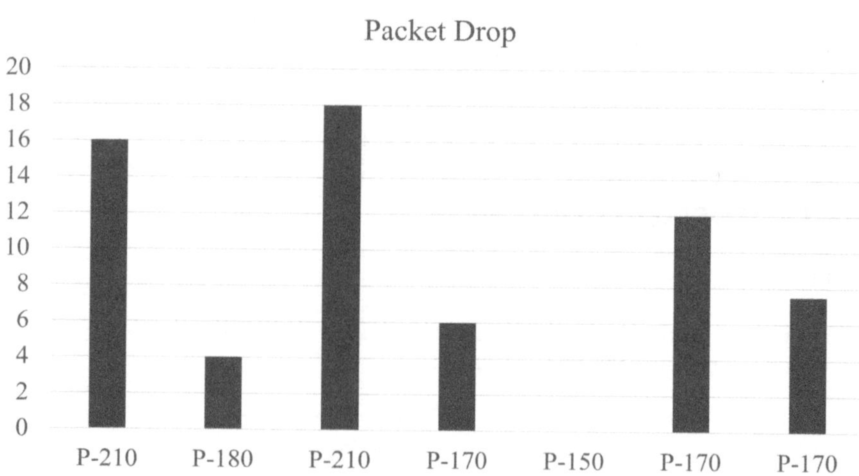

Figure 8.1 Packet drop scenario on round basis.

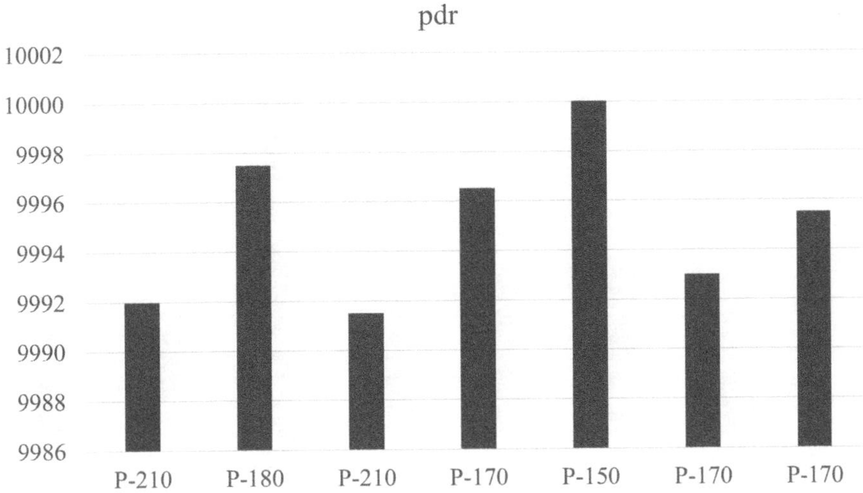

Figure 8.2 PDR scenario on round basis.

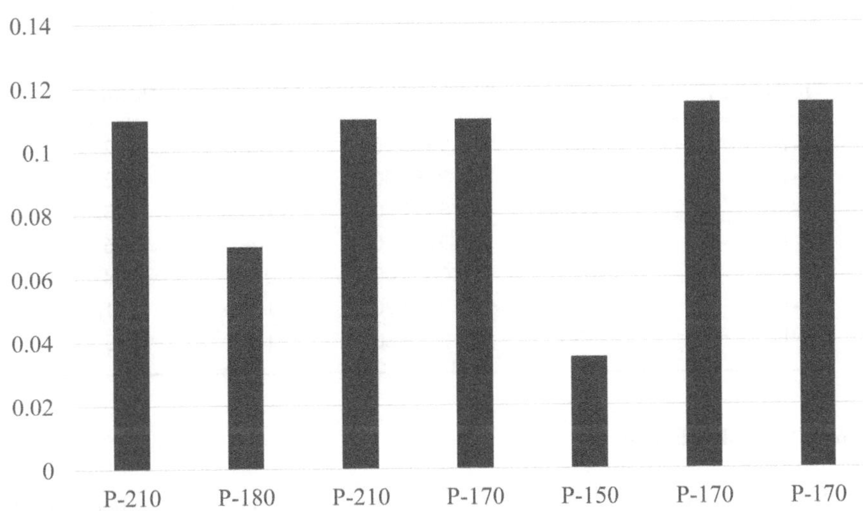

Figure 8.3 e2edelay scenario on round basis.

to the current sensor and is currently active when the question is received. Table 8.1 presents the results obtained from the execution of the WSN simulation, including Packet Drop, PDR, E2EDelay, and throughput.

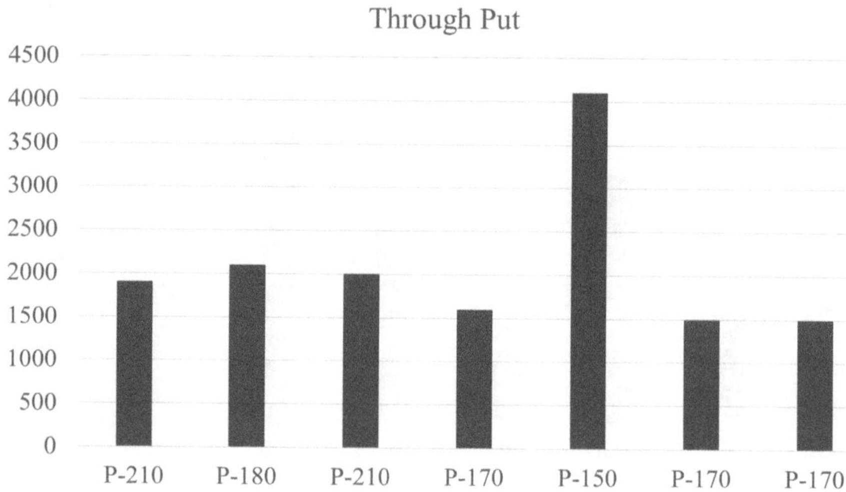

Figure 8.4 Through put scenario on round basis.

Table 8.1 Packet transmitted Packet Drop PDR, E2EDelay, and throughput

Packet transmitted	*Packet drop*	*PDR*	*E2EDelay*	*Throughput*
P-210	16	9992.10	0.1087	1932.10
P-180	4	9997.50	0.0684	2632.20
P-210	18	9991.40	0.1053	1993.40
P-170	6	9996.50	0.1065	1596.60
P-150	0	10000.00	0.0369	4064.90
P-170	12	9992.90	0.111	1531.40
P-170	7.5	9995.60	0.1109	1533.20

8.6 CONCLUSION

WSNs use anonymous, low-cost, low-power gadgets that are widely deployed to collect data from remote systems with minimal effort and human intervention. The use of sensors for the improvement of remote sensor systems was set by military applications, for example, in the combat zone area. In any case, remote sensor systems are now used in remote private applications, for example, environmental and regional testing, social insurance applications, home robotics, traffic control, and more. Security requests are a key task in many remote sensors. Due to unusual difficulties shown by the sensor, the use of security strategies in traditional systems cannot directly connect to the remote sensor system, according to their novel features. First of all, because the sensors' network consists of large sensor nodes, sensors are very sensitive to the cost of node manufacturers, and they should understand that the value of sensor nodes is less than $1 to make the sensor network possible. After that, the highest sensor keys could asset, memory, registration, and posts. Sensor nodes are generally powerful in batteries and in many cases, it is not possible to charge the battery. Maximum energy consumption is an important consideration in the largest sensor network protocols. Second, the sensor keys can be sent to unusual open areas, making them powerful against physical attacks by invaders. Usually, it is assumed that the attacker can control the sensor node outside the control and remove all secret data in the node. In addition, the sensor network is large sufficiently, the sensor nodes can be removed due to latency or failure, then the network topology is dynamically adjusted and the remaining nodes maintain the preferred functionality. It is possible to join the network. Finally, sensor systems use incredible remote compatibility channels and require frameworks. Therefore, existing security procedures are inadequate, and new procedures are needed. Large nodes can be very close to each other because many sensor nodes are centrally located. Therefore, multi-hop posts in the sensor system require less power than in general accounting. In addition, the transmission power level can be minimized, which is highly wanted in concealed operation.

Multi-hop posts can successfully kill a part of the mark effects experienced in remote posts. A stand between the most important obstacles to the sensors is low-power requirements. The sensor node is limited to a regular power supply. In this way, when traditional networks are designed to achieve high-quality service (Q) regulations, the sensor network protocol will have to pay attention to electricity savings primarily. End users need commercial closed mechanisms to enable network life cycles to increase the cost of low throughput and transmission. At present, many researchers are working on solutions to overcome these problems. Important features are unique in their flexibility in the WSN network format and sensor movement. This research describes the network, delays, and packet transmission rates. Jump transmission is performed using an H-F (hop field) Neural Network. The results of the rear-propulsion of our method compared with the HNN. The competition displays that the result of the jump field is better than the previous propagation. Here, the packet transfer rate and throughput increase, and the ending delays end up to the end. This chapter also describes how to effectively restore the wireless sensor network balance. In the future, neural-based and blur applications can be used to avoid clashes of data using these parameters.

REFERENCES

Agarwal, M., Gupta, V., Goel, A., & Dhiman, N. (2022). "Near lossless image compression using discrete cosine transformation and principal component analysis," *AIP Conference Proceedings*, Vol. 2481, No. 1, p. 020002. AIP Publishing LLC, Graphic Era University, Dehradun.

Baghyalakshmi, A., Ebenezer, J., & Satyamurty, S. A. V. (2011). "WSN based temperature monitoring for high-performance computing cluster," *2011 International Conference on Recent Trends in Information Technology (ICRTIT)*, Chennai, Tamil Nadu, pp. 1105–1110.

Bao, B., Qian, H., Wang, J., Xu, Q., Chen, M., Wu, H., & Yu, Y. (2017). Numerical analyses and experimental validations of coexisting multiple attractors in Hopfield neural network. *Nonlinear Dynamics*, 90(4), 2359–2369.

Cabeza, R. T., Viciedo, E. B., Prieto-Moreno, A., & Vega, V. M. (2016). Fault diagnosis with missing data based on Hopfield neural networks. In Antônio José da Silva Neto, Orestes Llanes Santiago, Geraldo Nunes Silva (eds), *Mathematical Modeling and Computational Intelligence in Engineering Applications* (pp. 37–46). Springer, Cham.

Cirstea, M., Cernaianu, M., & Gontean, A. (2012). "Packet loss analysis in wireless sensor networks routing protocols," *2012 35th International Conference on Telecommunications and Signal Processing (TSP)*, Prague, pp. 37–41.

Duan, S., Dong, Z., Hu, X., Wang, L., & Li, H. (2016). Small-world Hopfield neural networks with weight salience priority and memristor synapses for digit recognition. *Neural Computing and Applications*, 27(4), 837–844.

Gupta, V., & Seth, D. (2022). Design issues for developing routing protocols for flying ad hoc network. *Multimedia Technologies in the Internet of Things Environment*, 3(2022), 135–145.

Hu, S. G., Liu, Y., Liu, Z., Chen, T. P., Wang, J. J., Yu, Q., & Hosaka, S. (2015). Associative memory is realized by a reconfigurable memristive Hopfield neural network. *Nature Communications*, 6, 7522.

Patel, S. T., & Mistry, N. H. (2017). "A review: Sybil attack detection techniques in WSN," *2017 4th International Conference on Electronics and Communication Systems (ICECS)*, Coimbatore, pp. 184–188.

Rebentrost, P., Bromley, T. R., Weedbrook, C., & Lloyd, S. (2018). Quantum Hopfield neural network. *Physical Review A*, 98(4), 042308.

Wang, H., Yu, Y., Wen, G., Zhang, S., & Yu, J. (2015a). Global stability analysis of fractional-order Hopfield neural networks with time delay. *Neuro-computing*, 154, 15–23.

Xu, Q., Song, Z., Bao, H., Chen, M., & Bao, B. (2018). Two-neuron-based non-autonomous memristive Hopfield neural network: numerical analyses and hardware experiments. *AEU-International Journal of Electronics and Communications*, 96, 66–74.

Yadav, D. K. (2021). Application of IoT-Fog based real-time monitoring system for open-cast mines—A survey. *IET Wireless Sensor Systems*, 111, 1–21.

Yang, J., Wang, L., Wang, Y., & Guo, T. (2017). A novel memristive Hopfield neural network with application in associative memory. *Neuro-computing*, 227, 142–148.

Zhang, S., Yu, Y., & Wang, H. (2015). Mittag-Leffler stability offractional-order Hopfield neural networks. *Nonlinear Analysis: Hybrid Systems*, 16, 104–121.

Chapter 9

Computational approximation techniques for evaluating electromagnetic quantities

Meenu Kaushik and Vishant Gahlaut

9.1 INTRODUCTION

Electrostatics is basically a study of the interaction of point charges existing within a certain boundary. This phenomenon exists in a variety of applications, such as power transmission, solid-state electronic devices, vacuum tubes, and lighting equipment. The movement of charges is the basic concept behind the operation of these devices (Courant, 1943; Turner, 1956; Babuska, 1971; Zienkiewicz and Taylor, 2005; Babuska and Aziz, 1972a, 1972b). Various computer peripheral devices, such as magnetic memory, touchpads, liquid crystal displays, cathode ray tubes, and printers, utilize this concept as well. Electrostatics also plays a role in medical diagnosis and treatments, including the use of electrocardiograms and neurological examinations. Coulomb's law is used to calculate forces between charges, while Gauss's law is employed to estimate electric field intensity:

$$E = \frac{Q}{4\pi\epsilon_o R^2} a_R, \tag{9.1}$$

where, E is the electric field intensity, Q is the point charge, R is the distance between the point charge and point of interest where electric field intensity is to be calculated, ϵ_o is the permittivity of free space, and a_R is the unit vector pointing in the direction of force.

In practical scenarios involving multiple charges, the interaction between them is assessed using different computational techniques. Likewise, the presence of a magnetic field is ubiquitous. When an electric current flows through a conductor (whether it is a line, surface, or volume), a magnetic field is generated. Biot-Savart law and Ampere's law govern the existence of magnetic fields (Strang and Fix, 1973; Chen, 1984; Birdsall and Langdon, 1985; Oden, 1991; Lieberman and Lichtenberg, 1994). In the analysis of electromagnetic (EM) fields within a specific boundary, point-to-point calculations are required to estimate the overall effects of electric and magnetic fields. For different boundary conditions, proper structural analysis of the

DOI: 10.1201/9781003521389-9

geometries involved is necessary. In a homogeneous medium, EM quantities remain unchanged as they traverse through different materials. However, in a non-homogeneous medium where the fields cross two different media, discontinuities occur at the interface. At the boundary, the electric field has both normal and tangential components. The tangential component remains continuous, while the normal component experiences discontinuities. Similarly, the magnetic field has a continuous normal component and a discontinuous tangential component at the boundary, assuming that there is surface current flowing on this boundary. EM waves propagating in free space or vacuum, such as in wireless communication, are considered to travel through a homogeneous medium. Boundary conditions play a crucial role in solving EM problems. When an EM wave propagates through a boundary, it is partially transmitted across the boundary and partially reflected. Boundary equations are used to determine the direction and magnitudes of the incident, transmitted, and reflected EM waves. These equations are particularly important when computing EM fields in inhomogeneous media like cavities and waveguides. When an ideal conductor is considered, almost all incident waves on its surface are reflected, resulting in no field inside the conductor. At the surface of the conductor, specific conditions must be met. The magnetic field's normal component and the electric field's tangential component must both be zero. These conditions impose constraints on the patterns of EM fields when the medium is treated as a radio frequency (RF) cavity (Meenu Kaushik and Joshi, 2015; Kaushik and Joshi, 2015, 2016; Kaushik et al., 2021). These cavities serve as storage units for EM fields in the form of standing waves. The patterns and frequencies of these standing waves are analyzed based on the design of the cavity structure. The energy stored within the cavity is utilized to interact with and influence the particle beam in specific ways.

9.2 THE INDUCTIVE OUTPUT TUBE – A PROMISING RF VACUUM TUBE

The Inductive Output Tube (IOT) is a vacuum tube known for its ability to produce high RF power with high efficiency. It is widely used as a high-power source in communication transmitters and particle accelerators in the ultra-high-frequency (UHF) band. The IOT combines desirable properties of klystrons and tetrodes. It inherits the long life and better reliability characteristics of klystrons. It possesses the compactness and high efficiency traits of tetrodes. The IOT was invented in 1938. Figure 9.1 illustrates the schematic arrangement of the IOT, which consists of several major components. These include an electron gun, RF cavities, RF couplers, a collector, and a focusing magnet.

The operation of an IOT involves the use of an electron gun equipped with a control grid. The electron gun is used to emit a beam of electrons

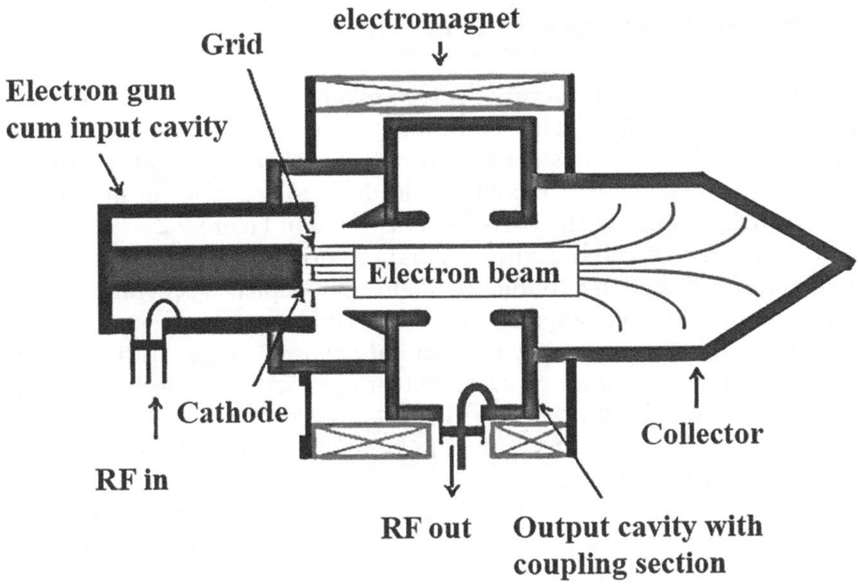

Figure 9.1 A schematic arrangement of IOT subassemblies.

under specific conditions. It consists of a cathode, control grid, anode, and beam focusing electrode (BFE). The input signal (typically a weak RF signal) is applied between the gap of the cathode and the control grid. This gap is referred to as the interaction gap. The presence of the input RF signal influences the emitted electrons. The velocities of electrons are varied within the interaction gap. The control grid is maintained at a more negative potential than the cathode potential. The control grid controls the flow of electrons. When no RF signal is present, it shuts off the electron beam. The variation in the velocities of electrons forms the bunches. These bunches get further accelerated by an anode maintained at a positive potential. The bunched beam is then directed toward the output section. This section is properly isolated from the input section of the IOT. Within the output section, the bunched beam transfers its kinetic energy to the weak, low-power input RF signal. This amplifies the RF signal to a higher level. The output cavity is equipped with an RF coupler. The amplified power is extracted from the tube with this arrangement. Due to the presence of numerous electrons in the beam, space-charge forces naturally cause the beam to spread. An electromagnet is employed to confine the beam in a straight path to mitigate this issue. This prevents collisions between the electrons within themselves and with the tube walls. It is crucial to minimize any wastage of electron beam energy to maintain the efficiency of the IOT. Therefore, a collector is employed in the IOT, which is responsible for dissipating the spent electron beam energy as heat (Meenu Kaushik and Joshi, 2015).

9.3 DESIGN APPROXIMATION OF A RADIAL RE-ENTRANT CAVITY FOR THE IOT

Re-entrant cavities play a crucial role in maintaining the resonance of electric and magnetic fields. They contribute to the amplification of RF power in the tube. These cavities are categorized as coaxial and radial structures depending on geometry. Coaxial cavities have a larger height in the z-axis and a smaller radius in the y-axis, as illustrated in Figure 9.2.

In contrast to coaxial cavities, radial re-entrant cavities are cylindrical structures with a center conductor. Figure 9.3 depicts a schematic view of this type of cavity. In radial cavities, the radial distance $(r_1 - r_0)$ is greater than the height of the cavity (h). Radial cavities are often the preferred choice for IOTs. They allow easier integration with the electron gun and reduce the complexities in the overall tube design. The radial cavity design provides sufficient space to accommodate the BFE. It serves as an outer envelope for confining the oscillating signal. This design enables the interaction between the electron beam and the input RF signal within the cavity. Various studies and publications have discussed the advantages and characteristics of radial cavities in the context of IOTs (Meenu Kaushik and Joshi, 2015; Meenu et al., 2016; Mercy et al., 2018; Vishant et al., 2021).

The electric and magnetic energies are stored within the cavity in the form of capacitance (C) and inductance (L). These parameters can be estimated

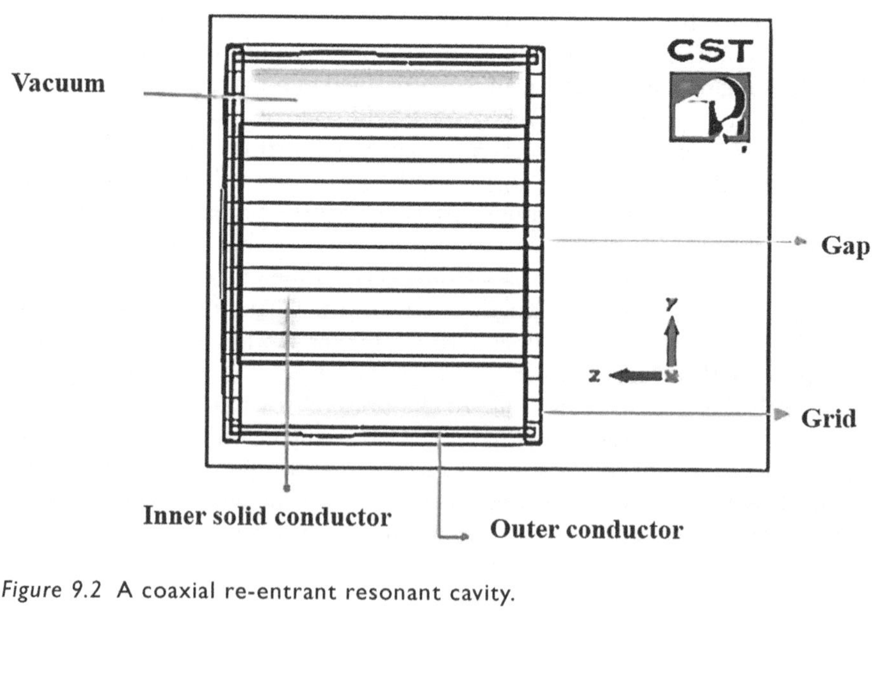

Figure 9.2 A coaxial re-entrant resonant cavity.

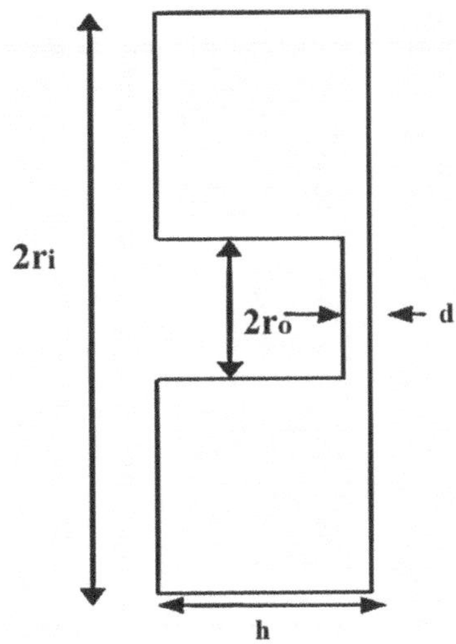

Figure 9.3 A schematic diagram of radial re-entrant cavity.

from the following equations (9.2) and (9.3), where r_1 and r_0 are the radii of outer and inner conductors, respectively, h is the height, d is the interaction space, μ_0 and ϵ_0 are the permeability and permittivity, respectively, in free space or vacuum.

$$L = \frac{\mu_0 h}{2\pi} \ln \frac{r_1}{r_0}, \tag{9.2}$$

$$C = \epsilon_0 \frac{\pi r_0^2}{d}, \tag{9.3}$$

The electric field is present within the interaction space (d), which is in the range of a few millimeters. When the electron beam reaches this space, there is an exchange of energies between the high-power electron beam and the low-power input RF signal. To perform EM analysis on resonant cavities, various tools such as SUPERFISH, MAGIC, and CST are commonly used. These tools are well-suited for eigenmode analysis of these cavity structures. They enable the determination of resonant frequencies. The electric field profiles generated within the interaction space can be observed. Studies and research papers have utilized tools like SUPERFISH code to calculate resonant frequency and electric fields. Figure 9.4 and 9.5 measure and visualize the resonant frequency and the electric field profile, respectively (Preist and Shrader, 1982; Shrader and Preist, 1984; POISSON/SUPERFISH Group of Codes, 1987; Kowalczyk, 2007; Larry Ludeking, 2011; McGinnis and

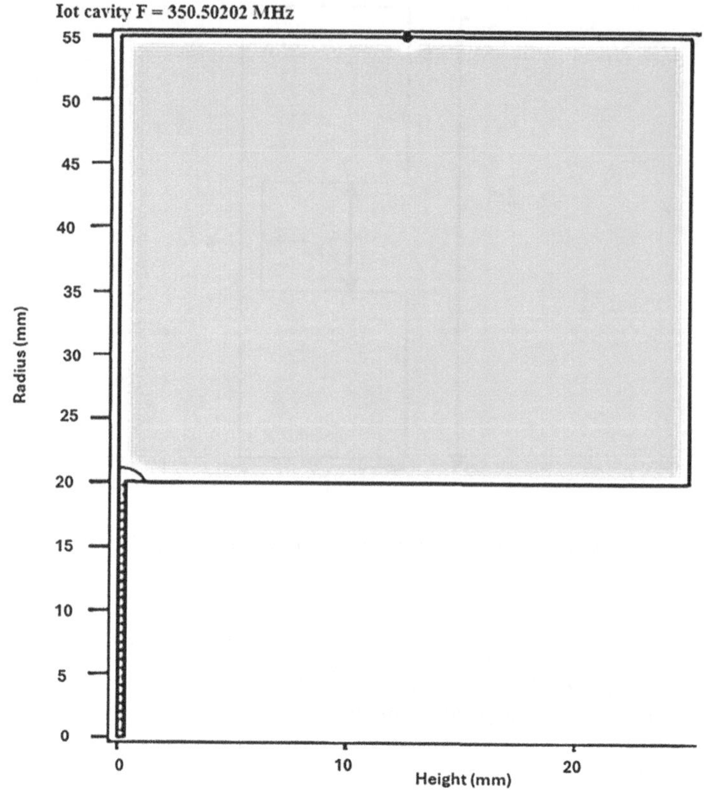

Figure 9.4 The resonant frequency and electric field obtained in SUPERFISH.

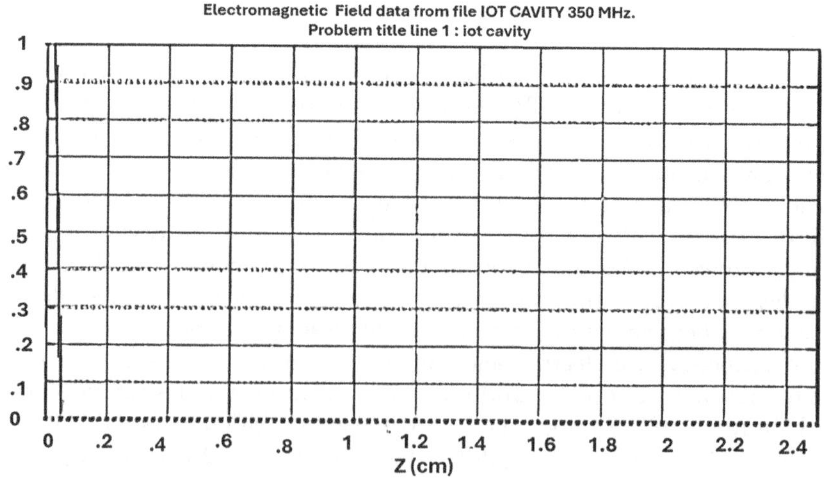

Figure 9.5 The plot of the electric field in the gap using the SUPERFISH code.

Lindroos, 2013; CST Studio Suite, 2016). Analytically, the resonant frequency of a narrow gap cavity is calculated using equations (9.4)–(9.6).

$$l_m = \frac{\sqrt{(r_1 - r_0)^2 + h^2}}{2}, \tag{9.4}$$

$$\lambda = 2\pi\sqrt{(r_0 h) * \left[\left\{\frac{r_0}{2d}\right\} + \left\{\frac{2}{\pi}\ln\frac{el_m}{d}\right\}\right] * \ln\frac{r_1}{r_0}}, \tag{9.5}$$

$$f = \frac{c}{\lambda}, \tag{9.6}$$

where the symbols have their usual meanings.

These software tools employ various approximation methods to compute eigenmode frequencies, EM fields, potentials, and other relevant quantities. These methods involve point-to-point analysis using well-defined equations. They are commonly utilized for solving EM problems. Numerous studies and publications have explored the application of these computational methods in solving EM problems in various fields. Some notable references include works by G. Borin et al. (2000); Marui and Martens (2002); Jackson (2003); M. L. Sisodia (2006); Pakarinen and Yeh (2009); Cohen and Helie (2009); Richard G. Carter (2010); Dempwolf and Zölzer (2011). The subsequent section will provide a detailed explanation of these computational methods.

9.4 COMPUTATIONAL APPROXIMATION METHODS INVOLVED IN COMPUTER-AIDED DESIGN

Various computational methods are employed in software tools for computer-aided design (CAD) to solve EM problems. The most commonly used methods are the finite difference method (FDM), the finite element method (FEM), and the particle in cell (PIC) method. The SUPERFISH group of codes utilizes both FDM and FEM approximations. It is primarily used to calculate electric and magnetic fields in static situations, to determine resonant frequencies, and to analyze field patterns in cavities. Cartesian and cylindrical coordinate systems are employed in this software. It is also used for designing electromagnets. The CST-MS (Computer Simulation Technology Microwave Studio) package utilizes FEM and PIC techniques for simulating different geometries. It is a comprehensive package designed for high-frequency EM analysis. The software provides various simulation solvers such as transient, frequency domain, and eigenmode analysis for computation. Hexahedral and tetrahedral meshes are used to discretize the geometry before conducting EM calculations. The MAGIC tool employs PIC simulations for particle tracking in different structures. It considers the interaction between space charge and EM fields in time domain. Users need to set up initial conditions. The simulations then perform EM point-to-point calculations as time progresses. The software ANSYS utilizes FEM for thermal and structural analyses of geometries. It is primarily a modeling package, which is extensively used for solving mechanical analysis. It is also capable of estimating thermal, structural, and fluid mechanics aspects for EM-related problems. Each of these software tools offers specific capabilities and features to tackle different aspects of EM analysis (ANSYS, n.d.).

9.4.1 FDM approximation

FDM is an elementary numerical method used to solve differential equations, including EM equations. It involves approximating a differential operator in both space and time domains. It divides the region into grids and labels them using a suitable coordinate system. This allows for the approximation of solutions at different points in space and time. The method utilizes known boundary conditions. It further uses the equations to estimate the difference between computed and expected solutions. FDM is a widely used approach for solving differential equations in various scientific and engineering fields. In one-dimensional (1D) problem, FDM can be approximated for a point x:

$$u'(x) = \lim_{0} h \frac{u(x+h) - u(x)}{h}, \tag{9.7}$$

Here, h should be very small to get an accurate approximation to the derivative or to minimize the truncation error. Suppose u is the function, which is C^2 continuous in the vicinity of x. For any $h > 0$,

$$u(x+h) = u(x) + hu'(x) + \frac{h^2}{2} u''(x + h_1), \tag{9.8}$$

where h_1 is a number between 0 and h. For finding out its solution, it is easier to consider only the first two terms, and the equation is modified as

$$u(x+h) = u(x) + hu'(x) + O(h)^2, \tag{9.9}$$

The term, $O(h^2)$, predicts that the approximation error is proportional to h^2. $C > 0$ is the constant, for $h > 0$ sufficiently small. The error introduced is of the order h. The approximation of u' is called the forward difference approximation. Hence, the FDM can be employed to solve potentials at different points within a defined region that is divided into small grids using (1D, 2D, 3D) rectangular coordinate systems.

For instance, let's consider a 2D region where the electric potential needs to be calculated. In this case, Maxwell's equations are utilized. Specifically, Laplace's equation is employed. Laplace's equation is defined as follows:

$$\nabla^2 V = 0, \tag{9.10}$$

Here, ∇^2 is called the Laplacian operator, which is expressed by equation (9.11):

$$\nabla^2 = \frac{\partial^2}{\partial x^2} + \frac{\partial^2}{\partial y^2} + \frac{\partial^2}{\partial z^2}, \tag{9.11}$$

This value of the Laplacian operator is valid for three-dimensional (3D) rectangular coordinate regions. The region grids are divided into (x, y) coordinates, and the potential is estimated at each and every grid point using the above FDM equations:

$$\text{As}\, u''(x) = \frac{u(x+2h) - 2u(x) + u(x-2h)}{4h^2}, \tag{9.12}$$

Here, function is the potential $V(x, y)$, therefore,

$$\frac{V(x+2h,y) - 2V(x,y) + V(x-2h,y)}{4h^2} + \frac{V(x,y+2h) - 2V(x,y) + V(x,y-2h)}{4h^2} = 0, \tag{9.13}$$

To estimate the electric potential, Laplace's equation needs to be solved. The electric field can be calculated by taking the gradient of the estimated potential (Autar Kaw et al., 2009). This involves calculating the partial derivatives of the potential with respect to the coordinates to determine the electric field components.

9.4.2 FEM approximation

Problems related to complex geometries and inhomogeneous media are computed using this method. The region of interest is divided into finite small elements. The method finds applications in the field of structural mechanics, fluid-heat transfer, and EM radiation. The problems associated with space and time dependence are expressed in partial differential equations (PDEs). For solving these equations, initial conditions must be provided. Poisson's equations, wave equations and Fourier laws can be solved using this technique. This method works on the principle of energy minimization. When specific boundary conditions are applied to a region, it can attain several configurations. The pattern having minimum energy is chosen as the solution. For initializing the estimation, four steps are executed.

The first step is to identify the relevant PDEs. These PDEs can be formulated in either differential or integral form, known as the strong and weak forms, respectively. Next, the region of interest is divided into smaller finite elements or a mesh, as shown in Figure 9.6. FEM can be applied to 1D, 2D, or 3D geometries, and the choice of element shapes depends on the specific problem.

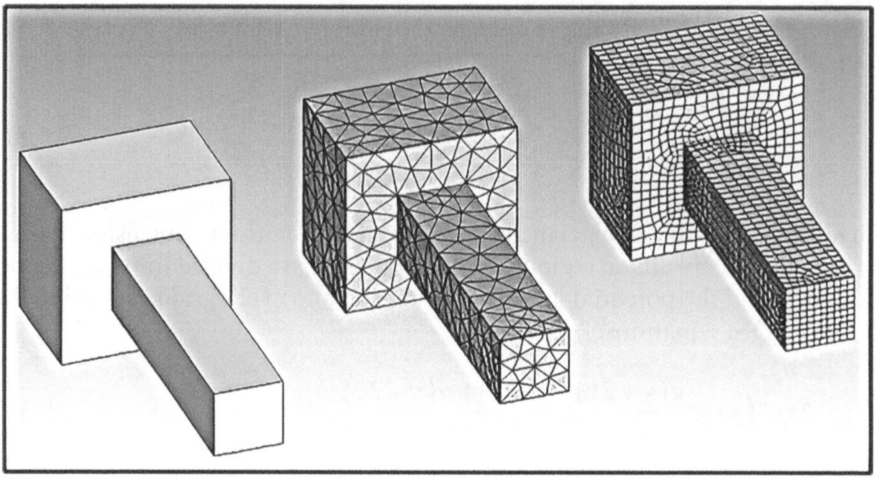

Figure 9.6 Discretization of geometry in meshes.

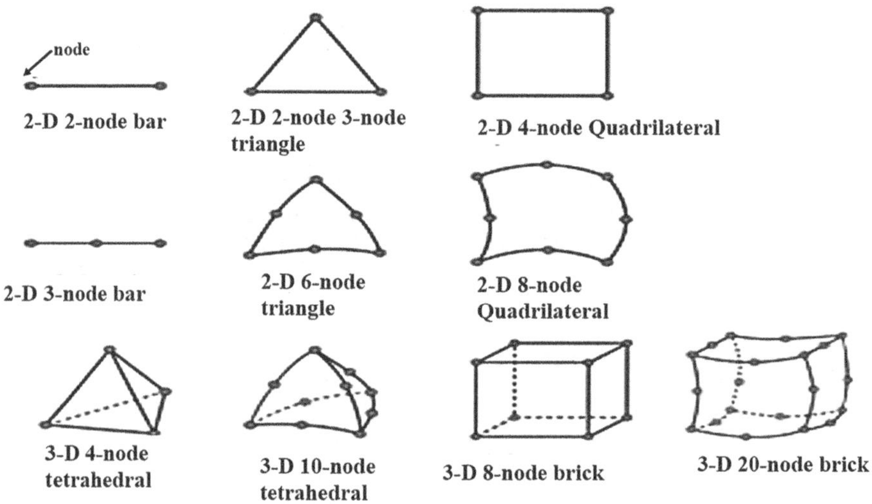

Figure 9.7 Types of 2D and 3D generated meshes.

Common element shapes include triangles, rectangles, quadrilaterals, tetrahedra, and hexahedra, as illustrated in Figure 9.7. Discretization involves transforming the integral form of the PDEs into a set of matrix equations, which can be solved using matrix algebra. This process converts the continuous problem into a discrete problem that can be solved numerically.

The FEM approximation process involves solving the governing equations for each individual element in the mesh and then combining the results to obtain the solution for the entire region of interest. Starting with a single

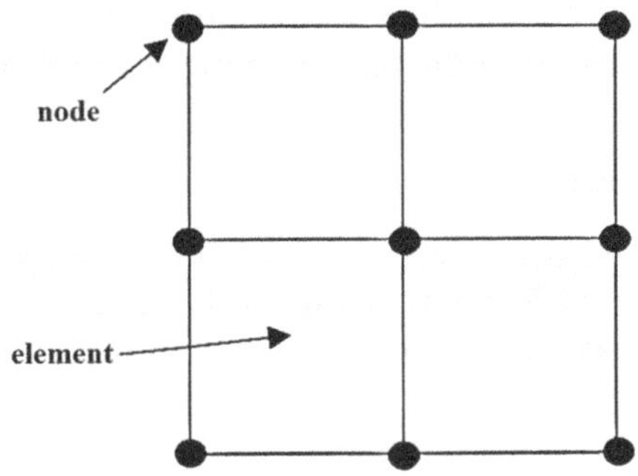

Figure 9.8 Nodes and elements in a subdivided grid.

element, the potential is calculated using appropriate coordinates within that element. Assuming the electric field to be uniform within the element, the gradient of the potential is then calculated using a matrix at the specified nodes of that element, as shown in Figure 9.8. This process allows for the estimation of the electric field within the element. By solving the equations for all elements in the mesh and assembling the results, a system of equations is obtained. These equations can then be solved to determine the solution for the entire region of interest. This approach allows for the computation of the electric potential and electric field throughout the entire domain by considering the contributions from each individual element.

In the FEM process, the following steps are typically followed: The geometry of the problem is created using appropriate CAD tools. The necessary geometric parameters are defined. The material properties for different regions or elements of the geometry are defined. This includes specifying electrical properties such as conductivity and permittivity. The initial conditions and boundary conditions are defined based on the problem requirements. These conditions specify the values or behavior of the solution at certain locations or boundaries. A mesh is generated to discretize the geometry into finite elements. The mesh consists of smaller elements such as triangles or rectangles in 2D, or tetrahedra or hexahedra in 3D, which approximate the shape of the geometry. The formulation of the elements involves defining the element type, shape functions, and interpolation functions that represent the behavior of the solution within each element. The governing PDEs are formulated in their weak form. This involves multiplying the PDEs by appropriate test functions and integrating over the domain. The global problem is set up by assembling the equations for each element into a system of equations that represents the entire problem domain. This is done by considering the connectivity between the nodes of adjacent elements. The system of equations is solved using numerical techniques such as matrix algebra or iterative methods to obtain the solution for the entire system. Post-processing analysis is performed to extract and visualize the desired quantities once the solution is obtained. The software like ANSYS implemented this method in the back-end programming (Moaveni, 2011).

9.4.3 PIC approximation

It is a technique used for computing the motion of the charged particles, electric potentials, and EM fields at various boundaries. The moving charges exert forces. Mathematically, Coulomb's law of force is expressed as:

$$\vec{F} = \frac{1}{4\pi\epsilon_0}\frac{Q_1 Q_2}{r^2} r_{12}, \tag{9.14}$$

Charged particles (or macro-particles) are represented by a finite number of computational particles. These macro-particles carry charge and mass, and their motion is governed by the equations of motion in the presence of EM fields. Rather than directly calculating Coulomb's force between individual particles, the PIC method solves the equations of motion for the macro-particles, taking into account the electric and magnetic fields at their respective positions. This method determines the Lorentz force acting on moving particles as follows:

$$\vec{F} = Q\left(\vec{E} + \vec{v} \times \vec{B}\right), \tag{9.15}$$

where Q is the particle charge, v is the particles' velocity, E and B are the electric and magnetic fields, respectively.

According to Newton's second law,

$$\frac{\overrightarrow{dx}}{dt} = \vec{v}, \tag{9.16}$$

$$\frac{\overrightarrow{dv}}{dt} = \frac{Q}{m}\vec{E}, \tag{9.17}$$

The electric field is then estimated using equation (9.18)

$$\vec{E} = -\nabla V, \tag{9.18}$$

where V is the electric potential, which was earlier estimated using Poisson's equation given by

$$\nabla^2 V = -\frac{\rho}{\epsilon_0}, \tag{9.19}$$

Here, ρ is the charge density, x is the displacement/position of particles, m and t are the particle mass and time, respectively.

In the PIC method, to accurately analyze the motion of particles, small-time steps are typically used in the simulations. This allows for capturing the dynamics of the particles and their interactions with EM fields more precisely. Several EM equations and principles are involved in the analysis, including Laplace's equation: It is solved to compute the electric potential distribution in the simulation domain. Equations of motion: The motion of particles is determined using Newton's laws of motion, taking into account the Lorentz force due to EM fields. Boltzmann relationship: It is used to estimate the electron number densities based on temperature and other parameters. Poisson's equation: It is solved to calculate the electric potential distribution based on charge densities. Electric field calculation: The gradient of the electric potential is computed to determine the electric field at each point in the simulation domain. Particle position and velocity estimation: The positions and velocities of the particles are updated based on the calculated forces and fields.

To analyze the motion of particles accurately, very small-time steps are chosen for simulations. EM equations are solved such as Laplace's equation, Newton's law, Boltzmann relationship, Poisson's equation, potential gradient, and the estimation of position and velocities of moving particles. Gauss elimination or the Gauss Seidel method is used for reaching the final solution. Specifically in microwave engineering, Euler's method and predictor corrector method are utilized to solve the Lorentz force equation for particle tracking in the presence of RF fields, space-charge fields, and magnetic fields. The PDEs and initial value and boundary value problems are computed for the desired results. The whole simulation runs in loops. The electric and magnetic fields are updated for newer time steps, and the corresponding parameters are analyzed iteratively. The region of interest is divided into grids. The charge density is computed by distributing the particle charges onto the computational cells' nodes in which particles are present. These are further divided into the node volume. For estimating the potential and electric fields, FDM is used in the discretized Poisson's equation. To analyze the motion of the particles, velocity is integrated over a time step. The position is updated again, and the velocity and electric fields at this current point in time are computed. New particles are generated using a suitable particle source. The expected outcomes are observed when the solution converges. Convergence is reached when the number of particles generated by the particle source becomes equal to the number of particles leaving the computational domain (Kahan, 1958; Birdsall and Langdon, 2005; Ren, 2011; Brieda, 2010, 2016).

9.5 RESULTS AND DISCUSSION

The resonant cavities are an integral part of a vacuum tube. They are employed to carry out optimum energy transfer from the electron beam to a weak RF signal. The detailed analysis of the resonant cavity is discussed. The operation of a vacuum tube is based on electromagnetism. Different electrical properties such as electric and magnetic fields, charges, potential, and energies come into play. The cavity structure must be designed to resonate to the operating frequency of the device. For proper designing of the cavity, all geometrical parameters are to be optimized. To achieve this, a combination of standard equations and optimization techniques is employed. The chapter provides a detailed explanation of this optimization process, presenting relevant equations and showcasing simulated results to illustrate the effectiveness of the design. By understanding and optimizing the dimensional parameters of the resonant cavity, it becomes possible to achieve the desired operating frequency and ensure efficient energy transfer within the vacuum tube. There are various simulation software options available for analyzing and estimating the design of different subassemblies within vacuum devices. The choice of software depends on the specific analysis requirements and desired outcomes. For simulating the eigenmode behavior of a cavity or resonating structure, where the focus is on frequency analysis, electric and magnetic field profiles, and surface current, software such as SUPERFISH, CST with eigenmode solver, and MAGIC can be considered. If the goal is to examine the thermal behavior of the structure, including temperature profiles in different regions and deformations caused by applied power, software like ANSYS and CST (for thermal analysis) can be used. To thoroughly analyze the electron beam within the operating vacuum tube, PIC analysis using software like CST particle tracking and MAGIC PIC analysis codes can be employed. It is important to select the appropriate software that best suits the specific simulation requirements and provides the necessary tools and capabilities to analyze and simulate the desired aspects of the vacuum device. The simulation codes mentioned utilize specific algorithms to compute complex problems. EM analysis is performed using FEM and FDM approximation techniques. On the other hand, when simulating the particle nature of the beam and its interactions with RF fields, PIC approximation methods are employed to obtain the desired results.

9.6 CONCLUSION

One of the applications of electromagnetism is in RF vacuum devices that are used as amplifiers to increase the low-power RF signal. This chapter focuses on the design and analytical study of the input cavity used in RF vacuum devices, specifically in the context of an IOT. The input cavity plays a crucial role as it not only handles the weak input RF signal but also houses the subassemblies of the electron gun. Its proper design is essential to ensure the desired operation of the IOT, which finds applications in signal broadcasting and particle accelerator communities. The chapter discusses the analytical modeling of resonant cavities, with a focus on radial re-entrant cavities that are commonly used in the IOT. While manual calculations can be performed for simpler geometries, complex structures require more time and effort, and the iterative nature of approximations makes manual analysis challenging and prone to human errors. To overcome these limitations, various software tools are explored, which offer faster solutions to EM problems using different mathematical approximation methods. The importance of approximation methods in predicting the operating conditions of microwave engineering devices has been discussed. It highlights the use of FDM for solving eigenmode frequencies, FEM for calculating EM fields and potentials in complex geometries by discretizing them into meshes, and PIC analysis for obtaining profiles and characteristics of charged particles in specific geometries.

REFERENCES

ANSYS. (n.d.) [online]. Ansys Academic Support | Tutorials & Training Materials.

Babuska, I. (1971). The finite element method for elliptic differential equations. In *Numerical solution of partial differential equations–II* (pp. 69–106). Academic Press.

Babuska, I. and Aziz, A. K. *Lectures on mathematical foundations of the Finite Element Method*, United States, N (1972a).

Babuska, I. and Aziz, A. K. Mathematical foundation of the finite element method with applications to partial differential equations. *Survey Lectures on the Mathematical Foundations of the Finite Element Method* (1972b) 3:636.

Birdsall, C. K. and Langdon, A. B. *Plasma Physics via Computer Simulation*, McGraw-Hill Publication, New York (1985).

Birdsall, C. K. and Langdon, A. B. *Plasma Physics via Computer Simulation*, Taylor & Francis Group, New York (2005).

Borin, G., De Poli, G. and Rocchesso, D. Elimination of delay-free loops in discrete-time models of nonlinear acoustic systems. *IEEE Transactions on Speech and Audio Processing* (2000) 8(5):597–604.

Brieda, L. The electrostatic particle in cell (ES-PIC) method. (2010) [online]. https://www.particleincell.com/2010/es-pic-method/

Brieda, L. Fundamentals of the particle in cell method. *Lecture materials* (2016). Particle in cell consulting LLC. California.

Carter, Richard G. Simple model of an inductive output tube. *IEEE Transactions on Electron Devices* (2010) 57(3).

Chen, F. F. *Introduction to Plasma Physics and Controlled Fusion*, Springer (1984).

Cohen and Helie, T. Simulation of a guitar amplifier stage for several triode models: examination of some relevant phenomena and choice of adapted numerical schemes. *127th Convention of Audio Engineering Society*, New York, United States (2009), 1–9.

Courant, R. Variational methods for the solution of problems of equilibrium and vibrations. *Bulletin of the American Mathematical Society* (1943) 49:1–23.

CST MICROWAVE STUDIO User Manual: AI Chat & PDF | Manualzz, 2016.

Dempwolf, K. and Zölzer, U. A Physically-Motivated Triode Model for Circuit Simulations. *14th International Conference on Digital Audio Effects*, Paris, France (2011), 257–264.

Jackson, D. L. Method and apparatus for the modeling and synthesis of harmonic distortion. *U.S. Patent No. 6,504,935 B1* (2003).

Kahan, W. Gauss-Seidel methods of solving large systems of linear equations. *Ph.D. thesis*, University of Toronto (1958).

Kaushik, M., Bandyopadhyay, A. K., Gahlaut, V. and Joshi, L. RF characterization of integrated gun-cavity for UHF band Inductive Output Tube. *22nd International Vacuum Electronics Conference (IEEE-IVEC)*, Rotterdam, Netherlands (2021), 1–2.

Kaushik, M. and Joshi, L. M. RF design of input cavity structure of a low frequency, high average power IOT. *PIERS Proceedings*, Prague, Czech Republic (2015), 850–853.

Kaushik, M. and Joshi, L. M. Design of spherical electron gun for ultra-high frequency, CW power inductive output tube. *AIP Conference Proceedings* (2016) 1715:020047.

Kaw, A., Nguyen, C. and Snyder, L. (2009). Finite difference Methods of Solving Ordinary Differential Equations. *Textbook notes of Finite Difference Methods of solving ordinary differential equations.*

Kowalczyk, R. D. Input circuit for a wideband IOT. *8th International Vacuum Electronics Conference (IEEE-IVEC)*, Kitakyushu, Japan (2007), 1–2.

Larry Ludeking. MAGIC 3.2.0 Help MANUAL Technical Report (2011) Alliant Techsystems (ATK) https://ia801803.us.archive.org/18/items/manualzilla-id-5975047/5975047.pdf

Lieberman, M. A. and Lichtenberg, A. J. Principles of plasma discharges and material processing, *MRS Bulletin* (1994) *30*(12), 899–901.

Marui, A. and Martens, W. L. Multidimensional perceptual calibration for distortion effects processing software. *Proceedings of the Audio Engineering Society 113rd Convention*, New York, Audio Engineering Society, ID:5708 (2002).

McGinnis, D. and Lindroos, M. The European spallation sources. *14th International Vacuum Electronics Conference (IEEE-IVEC)*, Paris, France (2013).

Meenu, K., Bandyopadhyay, A. K. and Joshi, L. M. Theoretical design of the gun-cum-input cavity and focusing structure envisaged for a 350 MHz, 100 kW (CW) inductive output tube. *Journal of Electromagnetic Waves and Applications* (2016) 30(14):1918–1928.

Meenu, K., and Joshi, L. M. Inductive output tube (IOT) – A review. *Journal of Electromagnetic Waves and Applications* (2015) 29(15):2027–2037.

Mercy, L. A., Gahlaut, V. and Ghosh, S. K. A model for quick thermal prediction of multi-stage depressed collector. *Journal of Electromagnetic Waves and Applications* (2018) 32(5):543–553.

Moaveni, S. *Finite Element Analysis – Theory and Application with ANSYS* (2011), 3/e. Pearson Education India.

Oden, J. T. *Finite Elements: An Introduction, Handbook: Numerical Analysis II, Finite Element Methods (Part I)*, North-Holland, Amsterdam (1991), 3–12.

Pakarinen, J. and Yeh, D. T. A review of digital techniques for modeling vacuum-tube guitar amplifiers. *Computer Music Journal* (2009) 33(2):85–100.

Preist, D. H. and Shrader, M. B. The Klystrode – An unusual transmitting tube with potential for UHF-TV. *IEEE Proceedings*, November 1982 70(11):1318–1325.

Reference manual for the POISSON/SUPERFISH Group of Codes, report, January 1, 1987; New Mexico. (https://digital.library.unt.edu/ark:/67531/metadc1317114/: accessed May 15, 2026), University of North Texas Libraries, UNT Digital Library, https://digital.library.unt.edu; crediting UNT Libraries Government Documents Department.

Ren, C. Introduction to particle-in-cell methods in plasma simulations. *HEDP Summer School* (2011) University of Rochester.

Shrader, M. B. and Preist, D. H. Klystrode update. *IEDM* (1984) 30.

Sisodia, M. L. *Microwave Active Devices: Vacuum and Solid State*. New Age International, New Delhi, India (2006), 3.47–3.49.

Strang, G. and Fix, G. J. *An analysis of the Finite Element Method*, Englewood Cliffs, Prentice-Hall (1973).

Turner, M. J., Clough, R. M., Martin, H. C. and Topp, L. J. Stiffness and deflection analysis of complex structures. *Journal of Aeronautical Science* (1956) 23:805–823.

Vishant, G., Lavi Deval, A. Latha, M. and Ghosh, S. K. Thermal study on the dependence of contact resistances on helical SWS of TWTs (2021).

Zienkiewicz, O. C., & Taylor, R. L. (2005). *The finite element method for solid and structural mechanics*. Elsevier.

Chapter 10

Analysis of AI-based emotion recognition techniques in online learning systems

Chandrakant Kumar Singh, Atul Sahu, and Ajender Kumar Malik

10.1 INTRODUCTION

The online learning systems have become an essential tool for the modern education system by creating opportunities and challenges for both learners and instructors. The COVID-19 pandemic has accelerated their use in schools and colleges for attending classes worldwide. The students and teachers had to depend on online digital platforms for teaching and learning tasks. This shift shows "how valuable online learning systems can be." Online learning systems helped both the teachers and students to connect through their mobile phones, laptops, and desktops, and they ensured continuity even during the pandemics or such global disruptions. Despite these benefits, they have also created several challenges. One of them was the difficulty the teacher faces in assessing students' engagement and emotional states or levels in virtual classrooms, which leads to reduced participation and poor learning outcomes. Likewise, in traditional classroom environments, the educators were able to observe "whether the student is understanding" or "not" by analyzing their "emotional states" and "feedbacks." With these feedbacks, the teacher adjusts their teaching strategies according to students' cognitive and emotional states.

But in virtual classrooms, since students and teachers were connected through their devices, teacher could not get such non-verbal feedbacks or clues, making it difficult to understand or recognize students' cognitive and emotional states. This made the teaching less effective, which created the need for such techniques that could help detect students' emotional states.

Using artificial intelligence, the emotion recognition has become an effective tool for overcoming these limitations. By analyzing learners' affective states through clues from facial emotion recognition (FER), speech- and text-based emotion recognitions, it is possible to identify the current emotional states of students, similar to the physical classroom environment. Thus, these approaches have significantly grown in the last few years. While using these approaches in online learning systems, the instructors were able to recognize the students' needs and their cognitive levels, which leads to good learning outcomes with a higher satisfaction level. These approaches

 DOI: 10.1201/9781003521389-10

have both their benefits and limitations when them used with different platforms. The main challenges include the technical limitations, privacy concerns, and cultural differences in recognition of the emotional states, which made the implementation more difficult.

In the last few years, several authors have proposed their approaches for better recognition of emotional states and implemented to the online learning environments. The authors have used several approaches, ranging from convolutional neural networks (CNNs) for facial expression recognition to recurrent and transformer-based architectures for speech- and text-based emotion detections. In addition, multimodal methods have been explored to combine complementary signals from different sources for better and reliable emotion recognition. While these studies highlighted the effectiveness of emotion-aware systems in education, there remains a need for an analysis considering the effectiveness, strengths, and limitations of these approaches in online learning environments. In this chapter, by addressing the limitations and strengths of these modalities for recognition of emotions, it provides suggestions for selecting appropriate emotion detection techniques for integrating into online learning environments. For the study, we consider four main emotion recognition modalities, such as facial expressions and speech-based and text-based emotion recognitions. In this chapter, you will find the applicability of such techniques in online learning environments, including Zoom, Microsoft Teams, and YouTube live streams. The main aim is to provide a clear guide and suggestions for the researchers and educators for the integration of emotion recognition techniques into their learning environments. Further, the study also highlights the need for adaptive pedagogical strategies while considering privacy and ethical guidelines.

10.2 RELATED WORKS

In recent years, several authors have published their work by highlighting the benefits and limitations of online learning systems. The authors have proposed their models while reporting higher accuracy using CNNs, LSTM, support vector machines (SVM), and multimodal emotion recognition approaches. For recognition of emotions, there are four main methods: text-based, speech-based, facial expressions, and multimodal-based emotion recognition techniques.

10.2.1 Text-based emotion recognition

Text-based emotion recognition has progressed significantly, with advances in deep learning and hybrid models. The text-based emotion recognition includes sentiments, emojis, and linguistic clues. The text-based emotion recognition analyzes the written contents, such as discussions, posts, chat messages, and essays. Canales and Martínez-Barco (2014) evaluated traditional ML methods such as SVM and Naive Bayes, achieving accuracy between 60% and 80%; sarcasm and multilingual issues were the challenges. Gaind et al. (2019) designed a hybrid NLP-ML system on Twitter and achieved 91.7% accuracy, while short and noisy texts were the challenges. Seyeditabari et al. (2018) reviewed the supervised and unsupervised lexicon methods; the supervised method reached 90% accuracy but struggled with metaphors and short texts. Acheampong et al. (2020) reviewed SVM and BERT models, achieved 70–85% accuracy, but sarcasm and multilingual data posed challenges. Thakur and Shrivastava (2018) reviewed bilingual emotion detection with SVM and achieved 80% accuracy, while slang or short text were the challenges. Li et al. (2023) deployed CNN-Random Forest with BERT embeddings on forum posts and achieved 85% accuracy, recommending multilingual real-time classification. Yasmina et al. (2016) used SVM for detecting and analyzing YouTube comments and achieved 70% accuracy, while informed language and emojis were the challenges. Chowanda et al. (2021) compared SVM, Naive Bayes, and LSTM for Twitter emotion recognition, with LSTM achieving 80% accuracy, while noisy and short texts were the main challenges. Kratzwald et al. (2018) used LSTM/CNN with BERT for business text, achieving 85% accuracy, and reported domain-specific language as the main challenge. Alwi et al. (2012) analyzed text chat in L2 learning, improved accuracy with a structured task, and reported challenges like off-task behaviors. Mu and He (2024) developed a VOCA-based virtual teacher with BERT sentiment analysis, which achieved 85% accuracy, while computational cost was the main challenge for such systems. In this text-based emotion recognition method, the authors reported that short and noisy text data, a lack of multilingual real-time classification, and sarcasm and multilingual data are the main challenges.

10.2.2 Speech-based emotion recognition

Speech-based emotion recognition (SER) is a type of emotion recognition via analyzing vocal features such as pitch, tone, rhythm, and intensity. SER uses deep learning models like CNNs and long short-term memory networks (LSTMs) for the extraction of meaningful patterns from audio signals. In order to recognize emotions in speech, machine learning and deep learning models have been used in combination with acoustic characteristics, including energy, pitch, and Mel-frequency cepstral coefficients (MFCCs). Vekkot et al. (2020) introduced DNN-PSO hybrids for emotional voice conversion, improving PESQ scores. CNN-LSTM architectures have shown consistent success with accuracy ranging from 70 to 90% on datasets like RAVDESS and IEMOCAP (Akçay & Oğuz, 2020; Er, 2020; Cheng et al., 2023; Begazo et al., 2024; Sareen & Seeja, 2025). However, challenges persist, including speaker variability, cross-corpus generalization, background noise, and the detection of subtle emotional cues (Athanaselis et al., 2005; Akçay & Oğuz, 2020; Er, 2020). These studies recommend data augmentation, attention mechanisms, particularly for applications in tutoring systems, virtual assistants, and real-time classroom monitoring (Zhao et al., 2024; Sareen & Seeja, 2025). The SER mainly used RAVDESS and IEMOCAP datasets and achieved 70–90% accuracy. SER is efficient for capturing tone and prosody in real time, while the noise and speaker variability are the main challenges.

10.2.3 Facial expression recognition

Recently, FER has seen rapid advancements with the adoption of CNNs, pre-trained architectures, and attention mechanisms. FER uses visual cues such as facial expressions, eye movements, and micro expressions to detect emotional states. Pereira et al. (2024) proposed vision-based FER with CNN/transformers, achieving 90% accuracy, and reported unconstrained settings as the main challenge. Abdullah and Alkan (2022) used ResNet18 for student engagement and reported 99% F1-score, while occlusion was the challenge. Aly (2025) integrated ResNet50 with CBAM/TCN modules and achieved 93% accuracy in online learning contexts. Fakhar et al. (2022) used VGG16 for real-time FER in the classroom, achieving 85–90% accuracy, and reported occlusions as the main challenge. Dukić and Krzic (2022) used ResNet-34 for real-time FER, achieving 85–90% accuracy and reported lighting variations as the challenge. Bartindale et al. (2021) developed ZoomSense with CNN-based FER, achieved 80% accuracy, and improved satisfaction, while API limitations were the challenges. Aly et al. (2023) used MobileNetV2 for FER, achieving 90% accuracy, while real-time processing was the challenge. While other studies (Khalfallah & Slama, 2015; Krithika & Lakshmi Priya, 2016; Petrovica et al., 2016; Yang et al., 2018; Rawal & Stock-Homburg, 2022; Cirneanu et al., 2023; ELsayed et al., 2023; Lawpanom et al., 2024; Singh et al., 2024) used CNN for FER, achieving 75–90% accuracy, with occlusions, reported real-time processing, and limited data reported as the main challenges. However, Ayvaz et al. (2017) used kNN/SVM for FER, achieving 85% accuracy, with real-time processing identified as the challenge. Kanematsu et al. (2016) linked blinking/temperature to engagement, while small datasets were the limitations.

10.2.4 Multimodal emotion recognition

Multimodal emotion recognition combines facial, speech, and textual modalities for enhancing the accuracies better compared to other unimodal approaches; however, synchronization and computation costs were reported as main challenges. Middya et al. (2022) employed CNN-LSTM architectures with audio-visual fusion, reporting 90% and 85% accuracies, respectively, but faced issues with data alignment and modality misalignment. In classroom contexts, Zhou and Liu (2022) applied CNN-LSTM and achieved 97% accuracy for anger detection, while noisy inputs were challenges. Similarly, Cheng et al. (2023) used a weight-adaptive CNN-LSTM for audio-video recognition, achieving 84% accuracy, but encountering modality imbalance was an issue. Pan et al. (2023) and Wu et al. (2025) highlighted hybrid fusion and transformer-based models and achieved 80–90% accuracy on datasets such as IEMOCAP, DEAP, and AVEC, while synchronization and cross-cultural variability remained challenging. Li (2024) demonstrated the potential of 2DCNN-LSTM in psychological counseling sessions with 92% accuracy, with privacy concerns being the main limitation. Similarly, Saxena et al. (2020) surveyed multimodal methods, noting CNNs achieved up to 99% accuracy in controlled settings but emphasized challenges in real-world cross-modal integration. These multimodal systems are highly robust, which combines multiple emotional signals for better emotion prediction, while synchronization, high computational costs, and privacy concerns were the main challenges.

10.2.5 Online learning platforms and student engagement

In education, AI is being used for learning, personalization, and making learning fast and effective. The online learning platforms have the potential to provide personalized learnings while maintaining student engagements. The online learning platforms are helpful for two contexts: the live video feeds and public livestreams using facial, speech-, and text-based emotion recognitions. Tajik (2025) showed Alexa–ChatGPT integration reduces learner anxiety and improved fluency, while Yang and Meinel (2014) enhanced lecture video retrieval system with ASR/OCR, while noise limited its performance. CNN-LSTM models by Naseer et al. (2024) boosted learner's engagement by 25%, but it raised privacy concerns. Khan et al. (2024) achieved 70% F1-scores in video summarization and suggested emotion integration on online platforms. Several surveys highlighted student's experiences: Salhab (2024) reported 60% accuracy but still preferred traditional classes, Zaim et al. (2024) found ethics agreements have raised participation by 15%, while Stephan et al. (2019) surveyed and reported higher anxiety in online courses. Hollister et al. (2022) reported 20% engagement drop, which was improved by interactive tools. The studies show the challenges, including stress in video conferencing (Okabe-Miyamoto et al., 2022), motivation and internet issues (Pratama et al., 2020; Yuzulia, 2021), depression in virtual learning (Azmi et al., 2022), and reported engagement gaps in Zoom/Meet contexts (Nehe, 2021; Oktaviani, 2021). Ethical and adaptive solutions were explored in a few studies and found that Zoom protocols improved communication (Nuary, 2022), flipped learning (Biju et al., 2020), and intelligent assistants (Casamayor et al., 2009). Broader reviews (Chassignol et al., 2018; Charitopoulos et al., 2020; Chen et al., 2020; Cheng et al., 2022; Wang et al., 2024) emphasized AI's potential for personalization, adaptive testing, and learning analytics while reporting challenges like scalability, bias, privacy, fairness, and accountability in real-world uses.

10.3 RESEARCH GAPS AND OUR CONTRIBUTIONS

Existing studies for supporting online learning through emotion detection have emerged after the COVID-19 pandemic. By analyzing text, speech, and facial expressions, the educators were now able to adjust their teaching methodologies based on students' emotional states. However, several key challenges still remain unresolved.

First, most of the researchers used existing or controlled datasets, which restricts the actual applicability in online learning settings, where variables like noise, illumination, cultural diversity, and masked face pose critical challenges. Second, few studies emphasize multimodal recognition, even though multimodal systems give more accurate results for modality imbalance and computational overhead. Third, many researchers have focused on emotion detection separately, determining how it could be implemented in the adaptive teaching frameworks. Lastly, despite their crucial role in online learning, ethical issues, such as students' permission, privacy, and data use, were frequently overlooked in these learning systems.

To bridge these gaps, this study provides a comparative analysis of existing methods by highlighting their suitability for education-specific applications. While proposing an integrated framework for integration of emotion detection, it also reports the need for ethical deployments in emotion-aware learning systems.

10.4 COMPARATIVE STUDY FRAMEWORK

Based on these studies, we conducted a comparative study for the evaluation of existing approaches across modalities (text, speech, facial, and multimodal) in terms of accuracy, datasets, challenges, and applicability to online learning environments.

10.4.1 Modality

Emotion recognition in online learning environments can be implemented in different ways, based on the type of modalities like text-based, speech-based, or facial expressions. The text-based methods analyze the written messages like chat, posts, or live comments, which is useful for platforms like YouTube or Facebook Live. The SER focuses on the way students/teachers speak or talk during sessions, extracting features like tone, pitch, and energy to understand their emotions, which works well for video calls on Zoom, Meet, and Microsoft Teams. The facial expression recognition finds the facial cues, like smiles, frowns, or blinking, to determine real-time students' emotions. Multimodal approaches combine two or more of these methods like text, speech, and facial expression recognitions, which helps more accurate emotion detection using information from multiple sources.

10.4.2 Techniques

For emotion recognition, several models have been trained and used across platforms. Earlier, SVM, Naive Bayes, and Random Forests were used as traditional machine learning methods for emotion recognition. With the advancements in deep learning, CNNs have become the popular choice for recognizing facial expressions, while LSTM networks and transformers are widely used for analyzing speech- and text-based emotion recognitions. Hybrid architectures such as CNN-LSTM combinations, attention-based models, and ensembles improved emotion recognition results by capturing patterns and analyzing over time and context. More recent research focuses on lightweight models such as MobileNetV2 and knowledge distillation, which are suitable for real-time use in online learning.

10.4.3 Datasets

Different types of emotion recognition models have been trained on well-known datasets and with custom-designed datasets. Text-based approaches have datasets such as ISEAR, SemEval, or social media sources, such as Twitter and Reddit. In speech-based approaches, they used datasets such as RAVDESS, IEMOCAP, Emo-DB, and classroom-specific recordings. For facial expressions, they used datasets like FER 2013, CK+, RAF-DB, AffectNet, or especially collected students' engagement data from online learning platforms. However, multimodal approaches combined text-based, speech-based, and facial expressions; the datasets like IEMOCAP, DEAP, AVEC, and custom classroom datasets were used for better fusion methods. These different datasets have different features and limitations, for example, FER 2013 datasets contain diverse images but is limited by low-quality and grayscale images, and ImageNet datasets include about 14 million diverse and high-quality images, but they require too complex modal architecture and computational costs.

10.4.4 Accuracy

The accuracy depends on the type of dataset and models used for emotion recognitions. The accuracy of text-based methods usually reaches 60–92% and is influenced by dataset quality and model design approaches. SER achieves 70–90%, while noisy environments and different datasets caused the performance drops. A modal achieving 85–90% accuracy can be considered ideal for text- and speech-based emotion recognitions. In facial expression recognition, the accuracy varies from 75 to 99% in controlled settings, which depends on the methodologies used, while diverse datasets are required for reliable emotion recognitions. The multimodal emotion recognition methods have outperformed unimodal systems, achieving 80–97% accuracy, making them the most accurate emotion recognition because these modalities process information from different sources and hence are able to give better emotion recognitions. However, 85–90% accuracy is typically expected from these modalities for better performance in the online learning systems.

10.4.5 Strengths

Each modality of emotion recognitions has its own strengths and benefits. Text-based methods are highly effective in discussion forums, chat systems, and public livestreams due to their ability to process large amount of data and hence provide scalable, asynchronous analysis. The SERs capture tone, pitch, and prosody from the audio to determine emotions that were hidden in text. This ability is really helpful for the online sessions and can be used to handle sarcasm, stress, or enthusiasm. Facial expression recognition determines facial visual cues such as smiles, frowns, or eye movements and provides real-time emotional state that can be used to manage learners' engagement, making it particularly powerful in online learning environments. Finally, multimodal systems are more robust, because they combines text, speech, and facial cues, to balance the weaknesses of individual modals and provide a more accurate students' emotion in online classroom environments.

10.4.6 Limitations

With the strengths, each modality has its limitations so it is wise to consider both while taking decisions for making effective changes. Text-based emotion recognition methods often miss sarcasm, hidden emotions, and struggles with multiple languages. Speech-based systems can be affected by background noise, different speaking styles, or cultural differences in voice were the limitations. Facial expression recognition is highly sensitive in recognizing emotions; it can be easily affected by poor lighting, low-quality cameras, occluded faces, and the use of masks, which makes it less reliable in such environments. While multimodal systems achieve better accuracy, they also face other issues or challenges, such as syncing different inputs, modality imbalance data, high computational costs, and privacy concerns. These limitations highlight the need for testing in real-world online classrooms and creating adaptive systems that can balance accuracy, efficiency, and ethical considerations.

To provide a clear overview, Table 10.1 gives the comparisons of these modalities while considering the modalities, their techniques, datasets, accuracy, and their strengths and limitations. While there are strengths and limitations in these modalities, a key question remains as to which one can be used reliably and effectively, because no single modality performs best for all tasks.

Table 10.1 Comparative study of emotion recognition modalities

Modality	*Techniques*	*Datasets*	*Accuracy (%)*	*Strengths*	*Limitations*
Text-based	SVM, NB, LSTM, BERT, Hybrid	ISEAR, SemEval	60–92	Works on chat/forums, context-aware	Fails on sarcasm, multilingual
Speech-based	MFCC+GMM, CNN-LSTM, DNN, Attn.	RAVDESS, IEMOCAP	70–90	Captures tone/prosody in real time	Noise, speaker variability
Facial (FER)	CNNs (VGG, ResNet), CNN-RNN	FER2013, AffectNet	75–99	Real time, visual engagement cues	Occlusion, lighting, masks
Multimodal	CNN-LSTM, Transformers, Hybrid	IEMOCAP, DEAP	80–97	Robust, combines signals	Sync issues, computation, privacy

Figure 10.1 shows the average accuracies of these modalities, including text-based, speech-based, facial (FER), and multimodal approaches. The text-based emotion recognition achieved 76% accuracy, SER achieved 80% average accuracy, FER achieved 87% average accuracy, and multimodal emotion recognition achieved 88.5% accuracies. These claimed accuracies have been calculated by taking the higher and lower limits and calculating their averages.

Figure 10.2 shows the comparative accuracy of the modalities from the 2018 to 2025 period, based on available literature. This study highlighted the progress of each modality in the past few years. For the study and comparison, we have calculated the average accuracies of that year. For example, if the accuracy range for a modality in a particular year is between 80 and 90%, the average will be 85%. Although taking the average of accuracies might not give precise accuracies, it offers a general view of trends and improvements over time.

Table 10.2 gives the comparison between different modalities with their respective accuracies. In this, we have taken year-wise comparisons of the modalities with their respective accuracies. Figures 10.1 and 10.2 were created using Python's matplotlib library, and the generated graphs are presented here as images. Figure 10.2 has been made from the data of Table 10.2.

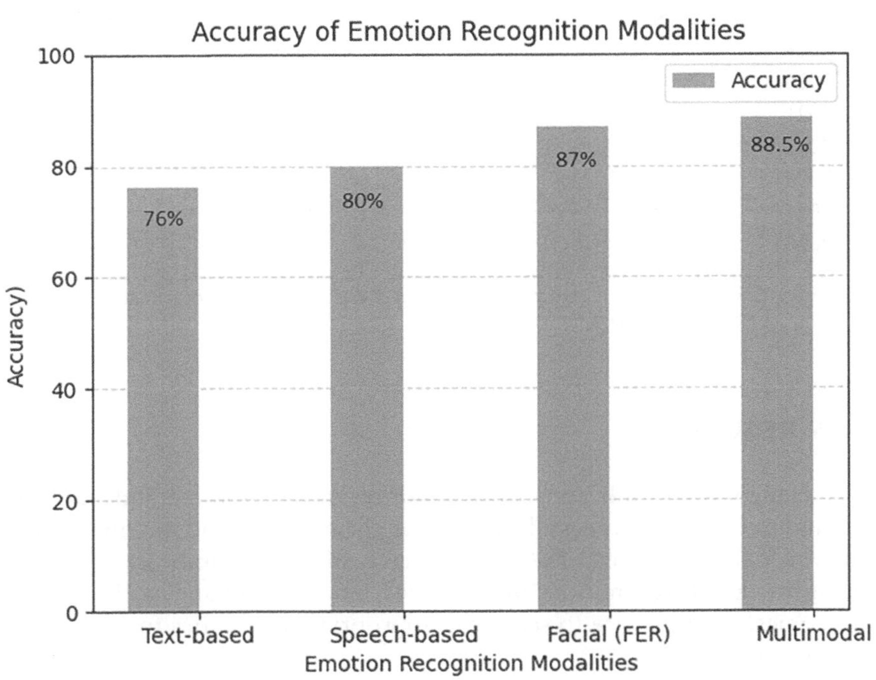

Figure 10.1 Average accuracy of emotion recognition modalities.

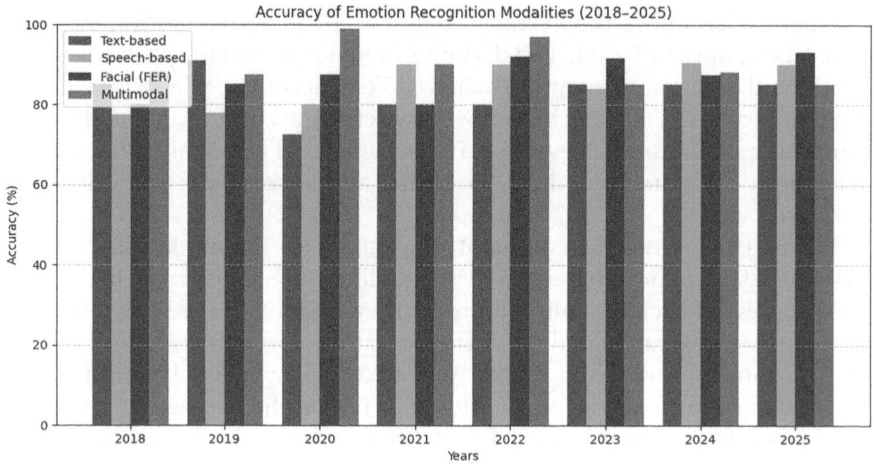

Figure 10.2 Average year-wise accuracy of emotion recognition modalities.

Table 10.2 Average year-wise accuracies of modalities

Years	*TER (%)*	*SER (%)*	*FER (%)*	*Multimodal (%)*
2018	85	77.5	80	87.5
2019	91	78	85	87.5
2020	72.5	80	87.5	99
2021	80	90	80	90
2022	80	90	92	97
2023	85	84	91.5	85
2024	85	90.5	87.5	88
2025	85	90	93	85

Note: FER, facial emotion recognition; SER, speech-based emotion recognition; TER, text-based emotion recognition.

10.5 RESULTS

The comparative study shows that emotion recognition methods have different levels of effectiveness depending on the modality and the environment in which they are used. The emotion recognition methods are much efficient for detecting emotions in online learning and enhancing the learning outcomes, but it is wise to select the appropriate modalities for the specific platforms.

10.5.1 Performance evaluation of the emotion detection modalities

In the text-based emotion recognition, it is suitable for asynchronous platforms such as YouTube/Facebook, where comments or live chat is used as a dataset for further work, achieving 60–92% accuracy, but the challenges are sarcasm, multilingual text, and implicit emotional cues; therefore, the models often struggle with the challenges. The text-based emotion recognition is good for chats and comments and would perform better if the limitations like sarcasm and multilingual issues were handled.

In the SER, CNN-LSTM hybrid approaches achieved 70–90% accuracy in synchronous platforms such as Google Meet, Zoom, and Microsoft Teams, where live video or audio feed is available. The challenges are low-quality audio, background noise, and speakers' variability due to cultural differences. The SERs are able to capture tone and stress.

In the facial expression recognition, it has been reported to have the highest accuracy in controlled environments, especially in the synchronous environment, where video feed was available, and achieved 75–90% accuracy, while ResNet variant reported 99% accuracy for real-time emotion monitoring. The challenges included occlusion, poor lightings, masked face, and low-quality video feed due to low internet speed. It is good for real-time visual clues and has the strength to perform well if the occlusions and lights are managed well.

In the multimodal emotion recognition, it performed better than other unimodal methods and achieved 80–97% accuracy by combining multiple signals across text, speech, and facial cues. The challenges include high computational costs, synchronization with modalities, and privacy concerns. These multimodal emotion recognitions are too good for integrating into the software that manages these computation and synchronization issues.

10.5.2 Recommendations for modalities selection

Based on this study, the selection of emotion recognition modal depends on the application and available resources for the online learning system. For asynchronous platforms like YouTube, Facebook live streams, and MOOCs, where only chat/comments are available, the text-based emotion recognition will be a correct practical option, due to its abilities to analyze large chat logs and comments. The text-based emotion recognition would perform better if its limitations, sarcasm, and implicit cues were considered. In synchronous environments, where the teachers and students connect together virtually for the act of teaching-learning process, including platforms, Zoom, Google Meet, Microsoft Teams, the speech-based and facial expression recognition methods are preferable. Due to its ability to capture real-time emotions, it can be used as a feedback system like informing the educators to adapt the teaching strategies as per students' emotions. In high-level environments, such as in universities or large communities, multimodal systems should be implemented because they integrate complementary signals and give more reliable performance, although these multimodal systems require higher computation costs. Conversely, for the settings where hardware and processing powers are limited, unimodal approaches like text, speech, and facial expression recognition still provide valuable insights for online learning systems while detecting emotions at a lower cost and with less computations.

10.5.3 Comparison between the strengths and limitations

The comparison between the strength and limitations of the modalities has been highlighted in Figure 10.3, and the graph was plotted using Python's matplotlib library based on the data of Tables 10.3 and 10.4. Table 10.3 shows the set of parameters for strength, and Table 10.4 shows the set of parameters for limitations, which helps clarify the selection of appropriate methods while considering its limitations.

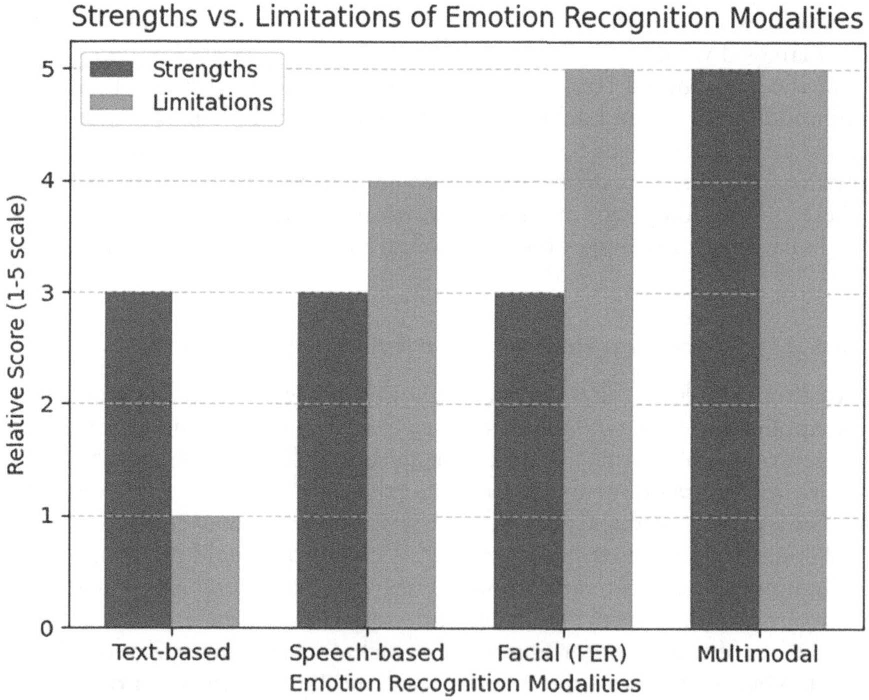

Figure 10.3 Comparison between strengths and limitations of emotion recognition modalities.

Table 10.3 Score calculations of strengths of modalities

Parameters	*Text-based*	*Speech-based*	*Facial (FER)*	*Multimodal*
Accuracy	Yes	Yes	Yes	Yes
Adaptability	Yes	Yes	Yes	Yes
Scalability	Yes	No	No	Yes
Real-time Applicability	No	Yes	Yes	Yes
Complementarity (multi-signal)	No	No	No	Yes

Table 10.4 Parameters for limitations of modalities

Parameters	*Text-based*	*Speech-based*	*Facial (FER)*	*Multimodal*
Noise/Data Sensitivity	No	Yes	Yes	Yes
Generalization Issue	Yes	Yes	Yes	Yes
Technical Considerations	No	No	Yes	Yes
Privacy/Ethics	No	Yes	Yes	Yes
Deployment Challenges	No	Yes	Yes	Yes

The score has been computed based on the strengths and limitations of the modalities. The parameters for the strengths of different modalities included features of adaptability to context (asynchronous/synchronous), scalability, real-time applicability, and ability to handle multiple signals. The parameters of strengths are given below:

1. Higher accuracy
2. Adaptability to context
3. Scalability to large-scale applications
4. Real-time applicability
5. Complementarity (multi-signal support)

The limitations of the modalities included the factors that affect the reliability, usability, and ethical considerations, like background noise, camera/mic quality, bandwidth, handling sarcasm and multilingual context, and lastly infrastructure demands. The parameters of limitations are given below:

1. Noise/Data Sensitive
2. Generalization Issues
3. Technical Considerations
4. Privacy/Ethics
5. Deployment Challenges

Table 10.5 Calculating the strengths and limitations of modalities

Modalities	*Strengths*	*Limitations*
Text-Based Emotion Recognition	3	1
Speech-Based Emotion Recognition	3	4
Facial Emotion Recognition	3	5
Multimodal Emotion Recognition	5	5

Figure 10.3 shows the scores obtained from the strengths and limitation sof the emotion recognition modalities. Table 10.5 shows the performance scores of the modalities based on the limitations and the strengths of emotion recognition modalities.

Overall, the result indicates that the multimodal emotion recognition provides the most reliable and robust performance for online learning systems if computational costs and synchronization issues were handled properly. This model yet has limitations when implemented in real-world educational platforms due to the challenges in combining different modalities. For lightweight implementations in online learning environments, the unimodal approaches such as speech or facial recognition still perform better for determining emotional clues and can be tailored for online learning platforms. Based on the requirements, these models can be used in online learning environments; however, their limitations must be considered, as they can perform similar to multimodal approaches.

10.6 FUTURE SCOPE AND RECOMMENDATIONS

This comparative analysis highlights several directions for implementing emotion recognitions in online learning systems. Although current studies show high accuracy in controlled datasets, future research must emphasize real-world classroom environments, for handling the challenges such as noise, cultural diversity, occlusions, and device variability, which significantly affect the performance. Large-scale, diverse, and ethically sourced classroom datasets would be essential for ensuring practical deployments and for the development of lightweight and real-time emotion detection with higher accuracies. While the multimodal systems consistently outperform unimodal approaches, their high computational cost requirements limit their adaptations in large-scale learning systems. Future work should focus on optimizing modals through compression and edge-computing strategies to make emotion recognition feasible for resource-constrained learning environments.

Ethical and privacy concerns must be considered while developing such emotion recognition integrated learning settings, because it involves sensitive data such as images, speech, and texts that could be used unethically. The future systems must integrate privacy-preserving techniques such as federated learning, anonymized data storage, and transparent consent mechanisms, and policies should clearly define data ownership and responsible AI practices to support their adoption in education systems.

Finally, the integration of an emotion recognition system into online learning gives a key opportunity to make the learner's engagement better. The models suggests the educators adapt learning strategies, whereas in the future, it should explore how affective states can dynamically guide teaching, thereby improving student engagements and supporting personalized learnings. Collaborative efforts should be made with AI researchers, educators, and policymakers for designing the system not only accurately but also considering ethical and privacy-aware systems.

10.7 CONCLUSION

The comparative study has analyzed the AI-based emotion recognition techniques used in online learning environments. Emotion recognition has significantly improved students' engagement and enhanced the overall accuracies. This study highlights the accuracies, strengths, and limitations of emotion detection modalities and proposes ethical and privacy-aware integration into online learning environments. The study also provides recommendations for the educators and policymakers while selecting appropriate modalities for their learning platforms. These findings indicate that the multimodal approaches often perform better, while unimodal methods remain suitable for lightweight system integration where real-time processings are required. However, challenges related to dataset diversity, real-time processing, and privacy remain unsolved. Overall, emotion-aware learning techniques hold significant potential to enhance the learning systems and guide future developments, while focusing on ethical, scalable, and pedagogically integrated solutions.

REFERENCES

Abdullah, M. U., & Alkan, A. (2022). A comparative approach for facial expression recognition in higher education using hybrid-deep learning from students' facial images. *Traitement Du Signal*, 39(6), 1929–1941. https://doi.org/10.18280/ts.390605

Acheampong, F. A., Wenyu, C., & Nunoo-Mensah, H. (2020). Text-based emotion detection: Advances, challenges, and opportunities. *Engineering Reports*, 2(7), e12189. https://doi.org/10.1002/eng2.12189

Akçay, M. B., & Oğuz, K. (2020). Speech emotion recognition: Emotional models, databases, features, preprocessing methods, supporting modalities, and classifiers. *Speech Communication*, 116, 56–76. https://doi.org/10.1016/j.specom.2019.12.001

Alwi, N. A. N. M., Adams, R., & Newton, J. (2012). Writing to learn via text chat: Task implementation and focus on form. *Journal of Second Language Writing*, 21(1), 23–39. https://doi.org/10.1016/J.JSLW.2011.12.001

Aly, M. (2025). Revolutionizing online education: Advanced facial expression recognition for real-time student progress tracking via deep learning model. *Multimedia Tools and Applications*, 84(13), 12575–12614. https://doi.org/10.1007/s11042-024-19392-5

Aly, M., Ghallab, A., & Fathi, I. S. (2023). Enhancing facial expression recognition system in online learning context using efficient deep learning model. *IEEE Access*, 11, 121419–121433. https://doi.org/10.1109/ACCESS.2023.3325407

Athanaselis, T., Bakamidis, S., Dologlou, I., Cowie, R., Douglas-Cowie, E., & Cox, C. (2005). ASR for emotional speech: Clarifying the issues and enhancing performance. *Neural Networks*, 18(4), 437–444. https://doi.org/10.1016/j.neunet.2005.03.008

Ayvaz, U., Gürüler, H., & Devrim, M. O. (2017). Use of facial emotion recognition in e-learning systems. *Information Technologies and Learning Tools*, 60(4), 95–104. https://doi.org/10.33407/ITLT.V60I4.1743

Azmi, F. M., Khan, H. N., & Azmi, A. M. (2022). The impact of virtual learning on students' educational behavior and pervasiveness of depression among university students due to the COVID-19 pandemic. *Globalization and Health*, 18(1). https://doi.org/10.1186/S12992-022-00863-Z

Bartindale, T., Chen, P., Marshall, H., Pozdniakov, S., & Richardson, D. (2021). ZoomSense: A Scalable Infrastructure for Augmenting Zoom. *MM 2021 - Proceedings of the 29th ACM International Conference on Multimedia*, 3771–3774. https://doi.org/10.1145/3474085.3478332

Begazo, R., Aguilera, A., Dongo, I., & Cardinale, Y. (2024). A combined CNN architecture for speech emotion recognition. *Sensors*, 24(17). https://doi.org/10.3390/S24175797

Biju, S. M., Salau, A. O., Eneh, J. N., Sochima, V. E., & Ozue, I. T. G. (2020). A novel pre-class learning content approach for the implementation of flipped classrooms. *International Journal of Advanced Computer Science and Applications*, 11(7), 131–136. https://doi.org/10.14569/IJACSA.2020.0110718

Canales, L., & Martínez-Barco, P. (2014). *Emotion Detection from Text: A Survey*. In *Proceedings of the workshop on natural language processing in the 5th information systems research working days (JISIC)* (pp. 37–43).

Casamayor, A., Amandi, A., & Campo, M. (2009). Intelligent assistance for teachers in collaborative e-learning environments. *Computers and Education*, 53(4), 1147–1154. https://doi.org/10.1016/J.COMPEDU.2009.05.025

Charitopoulos, A., Rangoussi, M., & Koulouriotis, D. (2020). On the use of soft computing methods in educational data mining and learning analytics research: A review of years 2010–2018. *International Journal of Artificial Intelligence in Education*, 30(3), 371–430. https://doi.org/10.1007/S40593-020-00200-8

Chassignol, M., Khoroshavin, A., Klimova, A., & Bilyatdinova, A. (2018). Artificial Intelligence trends in education: A narrative overview. *Procedia Computer Science*, 136, 16–24. https://doi.org/10.1016/J.PROCS.2018.08.233

Chen, L., Chen, P., & Lin, Z. (2020). Artificial intelligence in education: A review. *IEEE Access*, 8, 75264–75278. https://doi.org/10.1109/ACCESS.2020.2988510

Cheng, Y., Wu, G., Zou, H., Luo, P., & Cai, Z. (2022). A knowledge query network model based on rasch model embedding for personalized online learning. *Frontiers in Psychology*, 13. https://doi.org/10.3389/FPSYG.2022.846621

Cheng, Y., Zhou, D., Wang, S., & Wen, L. (2023). Emotion-recognition algorithm based on weight-adaptive thought of audio and video. *Electronics (Switzerland)*, 12(11). https://doi.org/10.3390/ELECTRONICS12112548

Chowanda, A., Sutoyo, R., Meiliana, & Tanachutiwat, S. (2021). Exploring text-based emotions recognition machine learning techniques on social media conversation. *Procedia Computer Science*, 179, 821–828. https://doi.org/10.1016/J.PROCS.2021.01.099

Cîrneanu, A. L., Popescu, D., & Iordache, D. (2023). New trends in emotion recognition using image analysis by neural networks. *Sensors*, 23(16), 1–12. https://doi.org/10.3390/s23167092

Dukić, D., & Krzic, A. S. (2022). Real-time facial expression recognition using deep learning with application in the active classroom environment. *Electronics*, 11(8), 1240. https://doi.org/10.3390/ELECTRONICS11081240

ELsayed, Y., ELSayed, A., & Abdou, M. A. (2023). An automatic improved facial expression recognition for masked faces. *Neural Computing and Applications*, 35(20), 14963–14972. https://doi.org/10.1007/S00521-023-08498-W

Er, M. B. (2020). A novel approach for classification of speech emotions based on deep and acoustic features. *IEEE Access*, 8, 221640–221653. https://doi.org/10.1109/ACCESS.2020.3043201

Fakhar, S., Baber, J., Bazai, S. U., Marjan, S., Jasinski, M., Jasinska, E., Chaudhry, M. U., Leonowicz, Z., & Hussain, S. (2022). Smart classroom monitoring using novel real-time facial expression recognition system. *Applied Sciences (Switzerland)*, 12(23). https://doi.org/10.3390/APP122312134

Gaind, B., Syal, V., & Padgalwar, S. (2019). Emotion detection and analysis on social media. *arXiv preprint arXiv:1901.08458*. https://arxiv.org/pdf/1901.08458

Hollister, B., Nair, P., Hill-Lindsay, S., & Chukoskie, L. (2022). Engagement in online learning: Student attitudes and behavior during COVID-19. *Frontiers in Education*, 7. https://doi.org/10.3389/FEDUC.2022.851019

Kanematsu, H., Ogawa, N., Shirai, T., Kawaguchi, M., Kobayashi, T., & Barry, D. M. (2016). Blinking eyes behaviors and face temperatures of students in YouTube lessons-for the future e-learning class. *Procedia Computer Science*, 96, 1619–1626. https://doi.org/10.1016/J.PROCS.2016.08.209

Khalfallah, J., & Slama, J. B. H. (2015). Facial expression recognition for intelligent tutoring systems in remote laboratories platform. *Procedia Computer Science*, 73, 274–281. https://doi.org/10.1016/J.PROCS.2015.12.030

Khan, H., Hussain, T., Ullah Khan, S., Ahmad Khan, Z., & Baik, S. W. (2024). Deep multi-scale pyramidal features network for supervised video summarization. *Expert Systems with Applications*, 237. https://doi.org/10.1016/J.ESWA.2023.121288

Kratzwald, B., Ilić, S., Kraus, M., Feuerriegel, S., & Prendinger, H. (2018). Deep learning for affective computing: Text-based emotion recognition in decision support. *Decision Support Systems*, 115, 24–35. https://doi.org/10.1016/J.DSS.2018.09.002

Krithika, L. B., & Lakshmi Priya, G. G. (2016). Student emotion recognition system (SERS) for e-learning improvement based on learner concentration metric. *Procedia Computer Science*, 85, 767–776. https://doi.org/10.1016/J.PROCS.2016.05.264

Lawpanom, R., Songpan, W., & Kaewyotha, J. (2024). Advancing facial expression recognition in online learning education using a homogeneous ensemble convolutional neural network approach. *Applied Sciences*, 14(3), 1156. https://doi.org/10.3390/APP14031156

Li, S., Deng, M., Shao, Z., Chen, X., & Zheng, Y. (2023). Automatic classification of interactive texts in online collaborative discussion based on multi-feature fusion. *Computers and Electrical Engineering*, 107, 108648. https://doi.org/10.1016/J.COMPELECENG.2023.108648

Li, X. (2024). Application of emotion recognition technology in psychological counseling for college students. *Journal of Intelligent Systems*, 33(1). https://doi.org/10.1515/JISYS-2023-0290

Middya, A. I., Nag, B., & Roy, S. (2022). Deep learning based multimodal emotion recognition using model-level fusion of audio–visual modalities. *Knowledge-Based Systems*, 244. https://doi.org/10.1016/J.KNOSYS.2022.108580

Mu, X., & He, J. (2024). Virtual teacher-aided learning system based on voice operated character animation. *Applied Sciences*, 14(18), 8177. https://doi.org/10.3390/APP14188177

Naseer, F., Khan, M. N., Tahir, M., Addas, A., & Aejaz, S. M. H. (2024). Integrating deep learning techniques for personalized learning pathways in higher education. *Heliyon*, 10(11). https://doi.org/10.1016/j.heliyon.2024.e32628

Nehe, B. M. (2021). Students' perception on google meet video conferencing platform during English speaking class in pandemic era. *English Review: Journal of English Education*, 10(1), 93–104. https://doi.org/10.25134/ERJEE.V10I1.5359

Nuary, R. H. (2022). Analysis of students learning ethics in online learning via zoom in mathematical communication ability at senior high school Bunda Kandung Jakarta. *Journal of Instructional Mathematics*, 3(1), 36–43. https://doi.org/10.37640/JIM.V3I1.1027

Okabe-Miyamoto, K., Durnell, E., Howell, R. T., & Zizi, M. (2022). Video conferencing during emergency distance learning impacted student emotions during COVID-19. *Computers in Human Behavior Reports*, 7. https://doi.org/10.1016/J.CHBR.2022.100199

Oktaviani, S. (2021). Online lectures using zoom application for undergraduate students during Covid-19 pandemic period. *Risenologi*, 6(1), 31–36. https://doi.org/10.47028/J.RISENOLOGI.2021.61.156

Pan, B., Hirota, K., Jia, Z., & Dai, Y. (2023). A review of multimodal emotion recognition from datasets, preprocessing, features, and fusion methods. *Neurocomputing*, 561. https://doi.org/10.1016/J.NEUCOM.2023.126866

Pereira, R., Mendes, C., Ribeiro, J., Ribeiro, R., Miragaia, R., Rodrigues, N., Costa, N., & Pereira, A. (2024). Systematic review of emotion detection with computer vision and deep learning. *Sensors*, 24, 3484. https://doi.org/10.3390/S24113484

Petrovica, S., Anohina-Naumeca, A., & Ekenel, H. K. (2016). Emotion recognition in affective tutoring systems: Collection of ground-truth data. *Procedia Computer Science*, 104, 437–444. https://doi.org/10.1016/J.PROCS.2017.01.157

Pratama, H., Azman, M. N. A., Kassymova, G. K., & Duisenbayeva, S. S. (2020). The trend in using online meeting applications for learning during the period of pandemic COVID-19: A literature review. *Journal of Innovation in Educational and Cultural Research*, 1(2), 58–68. https://doi.org/10.46843/JIECR.V1I2.15

Rawal, N., & Stock-Homburg, R. M. (2022). Facial emotion expressions in human-robot interaction: A survey. *Int J of Soc Robotics*, 14(7), 1583–1604. https://doi.org/10.1007/s12369-022-00867-0

Salhab, R. A. (2024). The zoom boom: Assessing videoconferencing attitudes among college students. *International Journal of Information and Education Technology*, 14(2), 204–213. https://doi.org/10.18178/ijiet.2024.14.2.2041

Sareen, V., & Seeja, K. R. (2025). Speech emotion recognition using Mel spectrogram and convolutional neural networks (CNN). *Procedia Computer Science*, 258, 3693–3702. https://doi.org/10.1016/j.procs.2025.04.624

Saxena, A., Khanna, A., & Gupta, D. (2020). Emotion recognition and detection methods: A comprehensive survey. *Journal of Artificial Intelligence and Systems*, 2(1), 53–79. https://doi.org/10.33969/AIS.2020.21005

Seyeditabari, A., Tabari, N., & Zadrozny, W. (2018). Emotion detection in text: A review. *arXiv preprint arXiv:1806.00674*. https://arxiv.org/pdf/1806.00674

Singh, J., Singh, A., Singh, K. K., Lal, B., Verma, H., Samudre, N., & Raperia, H. (2024). Real-time convolutional neural networks for emotion and gender classification. *Procedia Computer Science*, 235, 1429–1435. https://doi.org/10.1016/j.procs.2024.04.134

Stephan, M., Markus, S., & Gläser-Zikuda, M. (2019). Students' achievement emotions and online learning in teacher education. *Frontiers in Education*, 4. https://doi.org/10.3389/FEDUC.2019.00109

Tajik, A. (2025). Beyond voice recognition: Integrating Alexa's emotional intelligence and ChatGPT's language processing for EFL learners' development and anxiety reduction - A comparative analysis. https://doi.org/10.21203/rs.3.rs-5989702/v1

Thakur, P., & Shrivastava, R. (2018). A review on text based emotion recognition system. *International Journal of Advanced Trends in Computer Science and Engineering*, 7(5), 67–71. https://doi.org/10.30534/ijatcse/2018/01752018

Vekkot, S., Gupta, D., Zakariah, M., & Alotaibi, Y. A. (2020). Emotional voice conversion using a hybrid framework with speaker-adaptive DNN and particle-swarm-optimized neural network. *IEEE Access*, 8, 74627–74647. https://doi.org/10.1109/ACCESS.2020.2988781

Wang, S., Wang, F., Zhu, Z., Wang, J., Tran, T., & Du, Z. (2024). Artificial intelligence in education: A systematic literature review. *Expert Systems with Applications*, 252, 124167. https://doi.org/10.1016/J.ESWA.2024.124167

Wu, Y., Mi, Q., & Gao, T. (2025). A comprehensive review of multimodal emotion recognition: Techniques, challenges, and Future directions. *Biomimetics*, 10(7), 418. https://doi.org/10.3390/BIOMIMETICS10070418

Yang, D., Alsadoon, A., Prasad, P. W. C., Singh, A. K., & Elchouemi, A. (2018). An emotion recognition model based on facial recognition in virtual learning environment. *Procedia Computer Science*, 125, 2–10. https://doi.org/10.1016/J.PROCS.2017.12.003

Yang, H., & Meinel, C. (2014). Content based lecture video retrieval using speech and video text information. *IEEE Transactions on Learning Technologies*, 7(2), 142–154. https://doi.org/10.1109/TLT.2014.2307305

Yasmina, D., Hajar, M., & Hassan, A. M. (2016). Using YouTube comments for text-based emotion recognition. *Procedia Computer Science*, 83, 292–299. https://doi.org/10.1016/J.PROCS.2016.04.128

Yuzulia, I. (2021). The challenges of online learning during pandemic: Students' voice. *Wanastra: Jurnal Bahasa Dan Sastra*, 13(1), 08–12. https://doi.org/10.31294/W.V13I1.9759

Zaim, M., Zakiyah, M., & Zaim, R. A. (2024). Students' behavior in the learning process using zoom meeting media: Problems and solutions. *JOIV: International Journal on Informatics Visualization*, 8(3), 1168–1175. https://doi.org/10.62527/JOIV.8.3.2226

Zhao, G., Zhang, Y., & Chu, J. (2024). A multimodal teacher speech emotion recognition method in the smart classroom. *Internet of Things*, 25, 101069. https://doi.org/10.1016/J.IOT.2024.101069

Zhou, H., & Liu, Z. (2022). Realization of self-adaptive higher teaching management based upon expression and speech multimodal emotion recognition. *Frontiers in Psychology*, 13, 857924. https://doi.org/10.3389/fpsyg.2022.857924

Chapter 11

Approximation in the dynamics of chaotic maps using modified Mann orbit*

Ashish and Jinde Cao

11.1 INTRODUCTION

During the past two decades, chaos theory has significantly advanced the understanding of nonlinear dynamical systems by revealing fixed-point behavior, period-doubling transitions, and irregular dynamics. The concept of chaos was first described by H. Poincare [1] who examined the qualitative results in nonlinear dynamical systems. In 1960s, May [2] and Lorenz [3] examined more efficient and effective mathematical results, after that almost every field in science and engineering has been overflowed with experimental, as well as computational analysis in nonlinear systems. The difference and differential equations are considered the major backbone of the nonlinear dynamical systems in chaos theory. Therefore, V. F. Verhulst first introduced the population growth model using a nonlinear map known as a logistic equation. In 1978, the generic dynamical characteristics of the logistic equation were studied by the famous mathematician Feigenbaum [4]. After that, in a series of articles, the researchers and academicians published various articles, such as Devaney [5, 6], Holmgren [7], Andrecut [8], Ashish [9], and Alligood et al. [10].

Nowadays, the field of chaos theory is at the forefront of research in mathematics and is widely used in several scientific disciplines, including transportation problems, neural network systems, nonlinear inventory models, optimization, security, ecology, and epidemiological systems. Much of the interest in dynamics revolves around the chaotic behavior of a nonlinear system, which often is attained by using fixed-point iterative techniques. Song and Meng [11], using the error-valued iterative method, examined the synchronized dynamical behavior of the nonlinear system. Singh and Sinha [12] developed a communication system using a nonlinear dynamical system in 2013. After that, in 2013, Radwan [13] examined the dynamical

* Mathematical Subject Classification: 39A23, 39A28, 39A30

DOI: 10.1201/9781003521389-11

properties, such as fixed, periodic, and chaotic, for various modulated and generalized discrete difference maps. The stabilization in the aperiodicity of the recursive orbits of logistic-type discrete difference maps was determined by Parasad and Katiyar [14]. In 1953, a two-step feedback system in fixed-point theory was introduced by W. R. Mann [15], which gives superior dynamical properties in nonlinear systems. Chugh et al. [9] examined the convergence and stability in one-dimensional logistic maps in 2012. In 2019, Ashish et al. [16], using Mann's iterative scheme, examined the stabilization property in the standard logistic system, which gives extra freedom to control variables to stabilize a chaotic regime. In 2018, the chaotic behavior was studied in a superior orbit, and the applications in the transportation system were studied by Ashish et al. [17]. In 2014–2015, in a series of articles, Wu and Baleanu [18, 19] introduced the fractional order difference maps and then examined their dynamical properties through a bifurcation plot and Lyapunov exponent. Moreover, one may also read various latest articles on the dynamical properties of the nonlinear system, such as Ashish et al. [16, 20–23].

Recently, in 2022, Renu et al. [24] examined the dynamical characteristics of a novel discrete difference map using superior orbits such as periodicity, aperiodicity, and Lyapunov exponent. Further, the whole chapter is divided into four sections. In Section 11.1, the literature review on the dynamical characteristics and applications of the nonlinear dynamical systems is presented. In Section 11.2, basic properties and the construction of the modified dynamical system are studied. In Section 11.3, the experimental analysis is illustrated using a bifurcation plot in a modified Mann orbit, and finally, the chapter is concluded in Section 11.4.

11.2 NUMERICAL EVOLUTION IN A LOGISTIC-TYPE MAP VIA A MODIFIED MANN ORBIT

This section deals with the numerical as well as computational study of the logistic-type difference map using the modified Mann iterative technique. The dynamical behavior of the chaotic map exhibits sensitive dependence, period-doubling, aperiodicity, Lyapunov exponent, scaling of the orbits, and the complex behavior of nature. Therefore, this section determines the dynamical behavior of the logistic-type map using a two-step modified Mann iterative procedure. Thus, we consider the following discrete difference map:

$$f_r(x) = 1 - rx^2, \text{ where } x \in [-1,1] \text{ and } r \in [0, r_{\max}]. \tag{11.1}$$

Then, from the given equation (11.1), it is clear that the given system has two fixed points that belong to the closed interval [−1, 1]. Now, using the two-step modified Mann iterative procedure and the logistic-type equation (11.1), we obtain the following relation:

$$MM(r,\alpha,k,x) = \frac{(1-\alpha)}{k}x + \left(1 - \frac{(1-\alpha)}{k}\right)f_r(x)$$

$$MM(r,\alpha,k,x) = \frac{(1-\alpha)}{k}x + \left(1 - \frac{(1-\alpha)}{k}\right)(1 - rx^2), \qquad (11.2)$$

where $x \in [-1,1]$, $\alpha \in (0,1)$, $r \in [0, r_{\max}]$, and $k > 1$. It is noted that at $k = 1$, the given system (11.2) reduces to its original Mann iterative system and at $k = 1$ and $\alpha = 1$, it reduces to the trivial system $x_{n+1} = f(x_n)$. Therefore, we take $\alpha \in (0,1)$ and $k > 1$. Also, it is clear that the given system (11.2) depends upon the control parameters α, k and the growth-rate parameter r. For the sake of simplicity, let us take some particular values of the control parameter k and α. Also, the system (11.2) admits two fixed points:

$$x_1 = \frac{-1+\sqrt{1+4r}}{2r} \text{ and } x_1 = \frac{-1-\sqrt{1+4r}}{2r}.$$

11.3 EXPERIMENTAL EVOLUTION IN A LOGISTIC-TYPE MAP VIA A MODIFIED MANN ORBIT

The dynamical behavior of difference and differential equations is always affected by the growth-rate parameter r taken by the map and the corresponding control parameters. Therefore, this section deals with the dynamical properties of the discrete difference map (11.1) approximated by the modified Mann iterative procedure. The behavior is illustrated using the control parameters k, α, and the growth-rate parameter r. The results are examined for a few particular values of $\alpha = 0.9, 0.5, 0.1$ and $k = 2, 3, 5$, and the corresponding growth-rate parameter ranges are determined. Bifurcation plotting is shown using the superior modified Mann iterative system (11.2).

11.4 EVOLUTION OF DYNAMICAL BEHAVIOR WHEN $k = 2$ AND $\alpha = 0.9, 0.5, 0.1$

Taking the starter x_0 belongs to the closed interval $[-1,1]$, step size $h = 0.001$, $k = 2$, and $\alpha = 0.9, 0.5, 0.1$, the dynamical behavior is evaluated. For $k = 2$ and $\alpha = 0.9, 0.5, 0.1$, the growth-rate parameter admits $0 \le r \le 2.25$, $0 \le r \le 3.75$, and $0 \le r \le 7.25$, respectively. It is observed that as the value of the parameter α decreases from 1 to 0, the regime of the growth-rate parameter r increases rapidly, showing fixed state, periodic state, and the chaotic state of the system.

When we take $k = 2$ and $\alpha = 0.9$, the first periodicity in the system occurs at $r = 0.85$ through a modified Mann orbit, and after that the system fluctuates between two fixed point states of order 2, as shown in Figure 11.1. Similarly, the second bifurcation of order 4 is illustrated at $r = 1.41$, which means, for $1.41 < r \le 1.54$. Proceeding in this way, a sequence of periodic orbits of orders 2, 4, 8, 16, 32, 64, ... so on, for $1.54 < r \le 1.58$, is examined. Further, it is amazing to see that as the growth-rate parameter r approaches 1.58, the system admits a transition from periodicity to chaos. The completely chaotic nature of the system is shown in Figure 11.1 for $1.58 \le r \le 2.25$.

When $k = 2$ and $\alpha = 0.5$, the stability in fixed points occurs for $0 \le r \le 1.51$ and then the first bifurcation of order 2 takes place as r approaches to 1.51, as shown in Figure 11.2. Afterward, a sequence of periodic orbits of order 2, 4, 8, 16, 32, 64, ... is illustrated for the parameter range $1.51 < r \le 2.68$. As r approaches 2.68, an irregularity starts in the dynamical system, with a periodic window of order 3. Therefore, the chaotic regime occurs for the growth-rate parameter range $2.68 < r \le 3.75$, as shown in the bifurcation plot in Figure 11.2.

When $k = 2$ and $\alpha = 0.1$, the stable fixed-point state occurs for $0 \le r \le 3$ and then the first bifurcation of order 2 occurs as r approaches 3, as shown in Figure 11.3. Proceeding in this way, the periodic orbits of order 2^n are drawn for the growth-rate parameter $3 < r \le 5.2$. Afterward, as r approaches 5.2, an irregularity in the dynamical system (11.2) occurs up to 7.21. Also, a periodic window of order 3 is seen as r approaches 6.4. Therefore, the chaotic regime takes place for the growth-rate parameter range $5.2 < r \le 7.21$, as shown in the bifurcation plot in Figure 11.3.

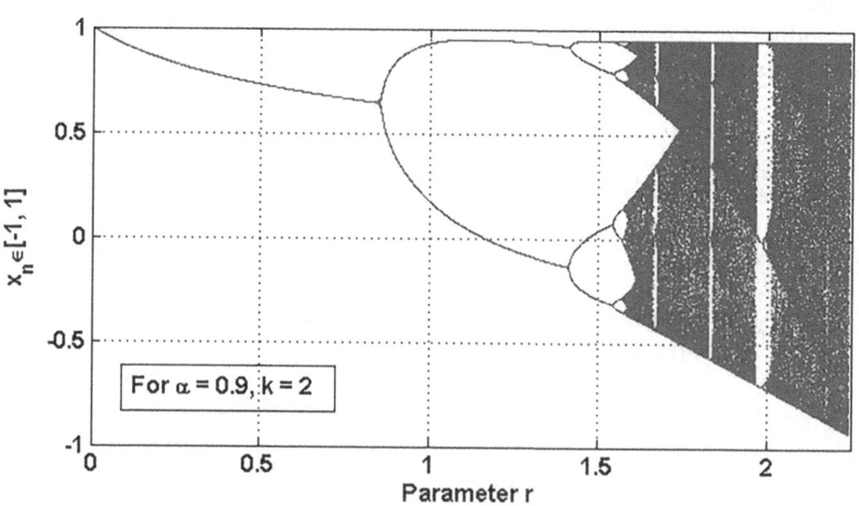

Figure 11.1 Bifurcation plot of the logistic-type map in the modified Mann orbit when $k = 2, \alpha = 0.9$, and $0 \le r \le 2.25$.

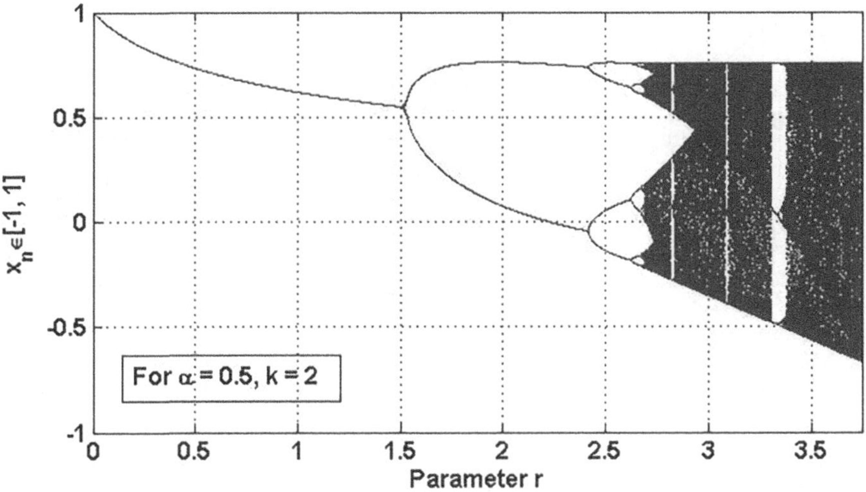

Figure 11.2 Bifurcation plot of the logistic-type map in the modified Mann orbit when $k = 2, \alpha = 0.5$ and $0 \le r \le 3.75$.

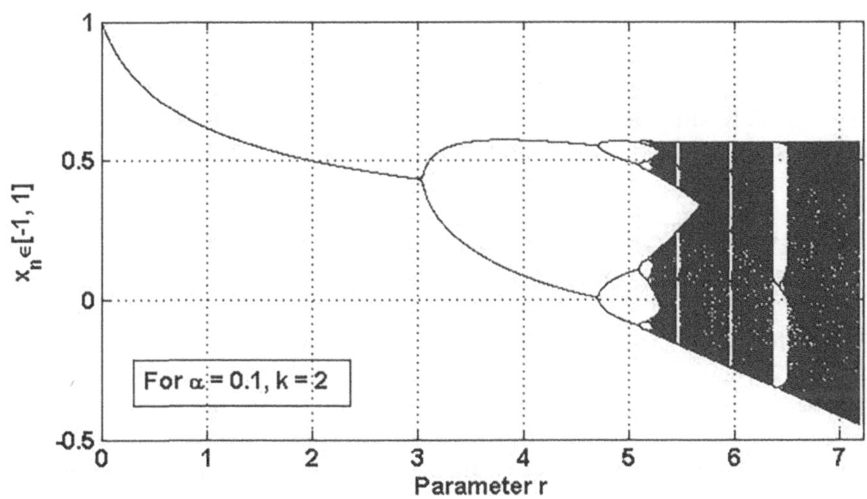

Figure 11.3 Bifurcation plot of the logistic-type map in the modified Mann orbit when $k = 2, \alpha = 0.1$, and $0 \le r \le 7.21$.

Figures 11.1–11.3 show the complete dynamical representation of the logistic-type map in the modified Mann orbit for the particular values of the parameters $k = 2$ and $\alpha = 0.9, 0.5, 0.1$. Throughout the study, it is examined that as the parameter k is fixed at 2 and the modified parameter α is decreased from 0.9 to 0.1, the growth-rate parameter r range increases

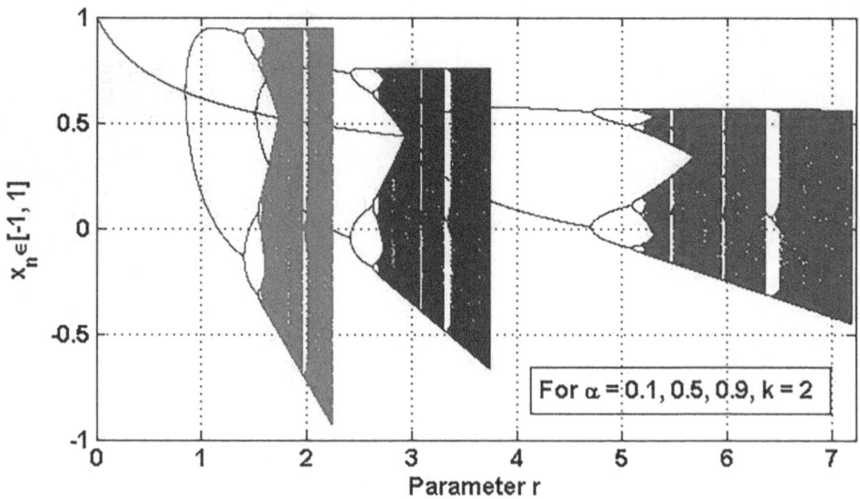

Figure 11.4 Comparative analysis versus bifurcation plots of the logistic-type map in the modified Mann orbit when $k = 2$ and $\alpha = 0.9, 0.5, 0.1$.

simultaneously, as shown in Figure 11.4. That means, for $\alpha = 0.9$, it admits $0 \le r \le 2.25$, for $\alpha = 0.5$, it admits $0 \le r \le 3.75$, and for $\alpha = 0.1$, it admits $0 \le r \le 7.21$. Amazingly, all the bifurcation plots illustrate stable fixed-point states, stable periodic states, and chaotic states.

11.5 EVOLUTION OF DYNAMICAL BEHAVIOR WHEN $k = 3$ AND $\alpha = 0.9, 0.5, 0.1$

Here, for $k = 3$ and $\alpha = 0.9, 0.5, 0.1$, the growth-rate parameter admits $0 \le r \le 2.16, 0 \le r \le 3$, and $0 \le r \le 4.36$, respectively. When $k = 3$ and $\alpha = 0.9$, the first bifurcation occurs at $r = 0.8$ in the modified Mann orbit (11.2), and after that, the system alternates between two periodic states of order 2, as shown in Figure 11.5. Similarly, the second bifurcation of order 4 is illustrated at $r \approx 1.35$, which means, for $1.35 < r \le 1.48$. Then, the higher order periodicity for order 2^n exists for $1.48 \le r \le 1.514$. Finally, as $r \approx 1.514$, the recursive orbit approaches the chaotic regime for $1.514 < r \le 2.16$, followed by a periodic window of order 3 at $r \approx 1.9$.

As discussed earlier, when we take $k = 3$ and $\alpha = 0.5$, the fixed-point stability exists for $0 \le r \le 1.17$ and the periodicity of order 2^n exists for $1.17 < r \le 2.12$. After that, the system starts to vibrate irregularly, that is, chaos occurs as r approaches 2.12. The complete chaotic behavior occurs for $2.12 < r \le 3$, followed by a periodic window of order 3, as shown in the bifurcation in Figure 11.6.

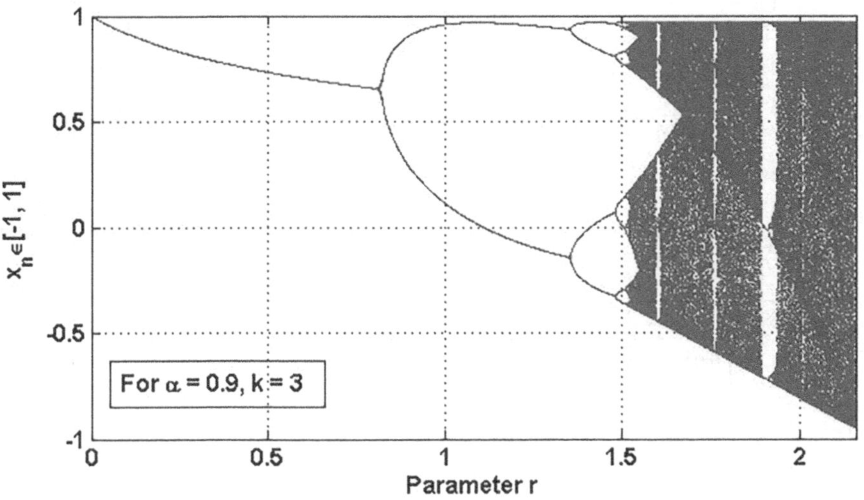

Figure 11.5 Bifurcation plot of the logistic-type map in the modified Mann orbit when $k = 3, \alpha = 0.9$, and $0 \le r \le 2.16$.

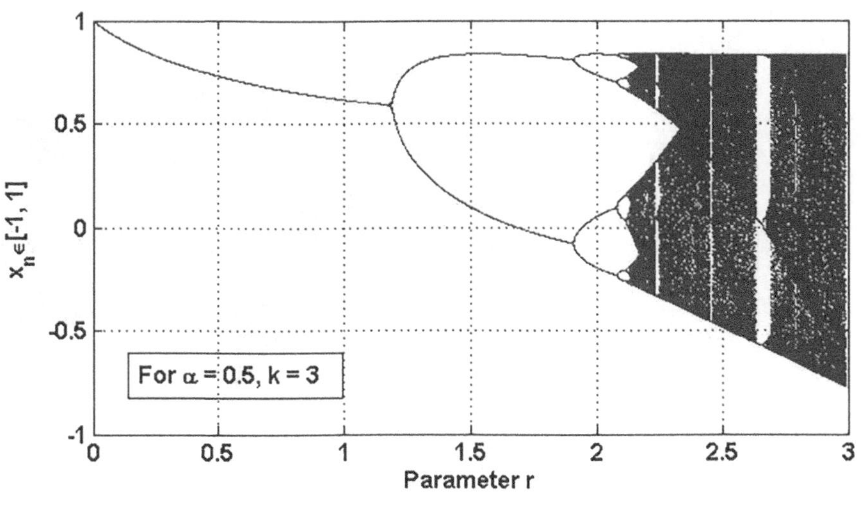

Figure 11.6 Bifurcation plot of the logistic-type map in the modified Mann orbit when $k = 3, \alpha = 0.5$ and $0 \le r \le 3$.

When $k = 3$ and $\alpha = 0.1$, the fixed-point stability exists for $0 \le r \le 1.75$ and the periodicity of order 2^n exists for $1.75 < r \le 3.12$. After that, the system starts to vibrate irregularly, that is, chaos occurs as r approaches 3.12. The complete chaotic behavior occurs for $3.12 < r \le 4.36$ followed by a periodic window of order 3 at $r = 3.85$, as shown in the bifurcation in Figure 11.7.

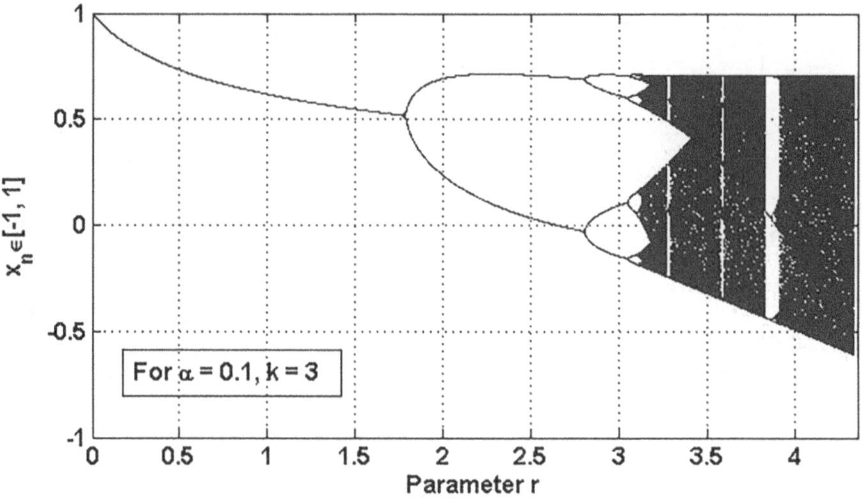

Figure 11.7 Bifurcation plot of the logistic-type map in the modified Mann orbit when $k = 3, \alpha = 0.1$, and $0 \leq r \leq 4.36$.

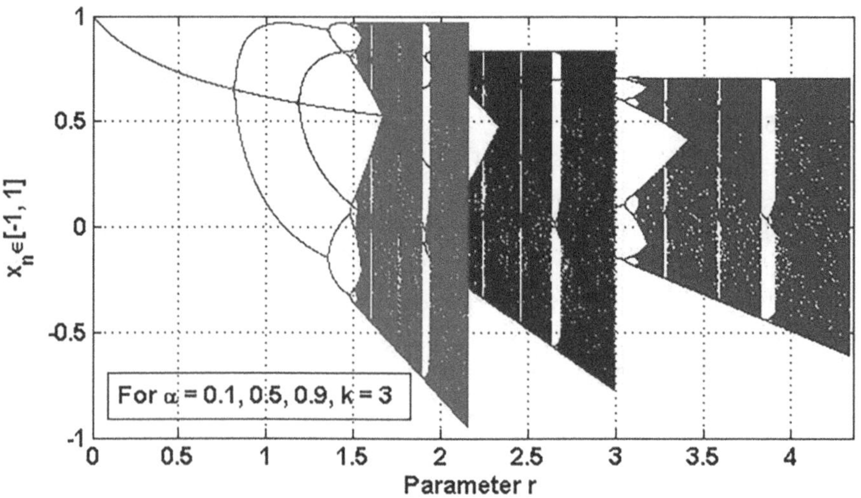

Figure 11.8 Comparative analysis versus bifurcation plots of the logistic-type map in the modified Mann orbit when $k = 3$ and $\alpha = 0.9, 0.5, 0.1$.

From Figures 11.5–11.7 and the comparison in Figure 11.8, the following analysis is carried out for the logistic-type map using the modified Mann orbit:

- For $k = 3$ and $\alpha = 0.1$, the complete bifurcation occurs for $0 \leq r \leq 4.36$.
- For $k = 3$ and $\alpha = 0.5$, the complete bifurcation occurs for $0 \leq r \leq 3$.

- For $k = 3$ and $\alpha = 0.9$, the complete bifurcation occurs for $0 \le r \le 2.16$.
- As α decreases from 1 to 0, the growth-rate parameter increases continuously.
- The range of the fixed, periodic, and chaotic states also increases simultaneously.

11.6 EVOLUTION OF DYNAMICAL BEHAVIOR WHEN $k = 5$ AND $\alpha = 0.9, 0.5, 0.1$

In this section, for $k = 5$, and $\alpha = 0.9$, the complete bifurcation takes place for $0 \le r \le 2.1$. Then, the bifurcation plot is divided into three stages: fixed, periodic, and chaotic. In the first stage, the fixed-point state occurs for $0 \le r \le 0.78$, in the second stage, the periodicity occurs for $0.78 < r \le 1.47$, and finally, the chaotic state takes place for $1.47 < r \le 2.1$. Figure 11.9 shows the complete dynamical nature of the given system.

When $k = 5$ and $\alpha = 0.5$, the fixed-point stability exists for $0 \le r \le 0.97$ and the periodicity of order 2^n exists for $0.97 < r \le 1.787$. After that, the system starts to vibrate irregularly as r approaches 1.787, which means, chaos occurs. The complete chaotic behavior occurs for $1.787 < r \le 2.5$, followed by a periodic window of order 3 at $r = 2.22$, as shown in the bifurcation in Figure 11.10.

Figure 11.11 shows the complete bifurcation plot of the logistic-type map using a modified Mann iterative procedure for the control parameter $k = 5$ and $\alpha = 0.1$. It shows the largest bifurcation plot for the full range of

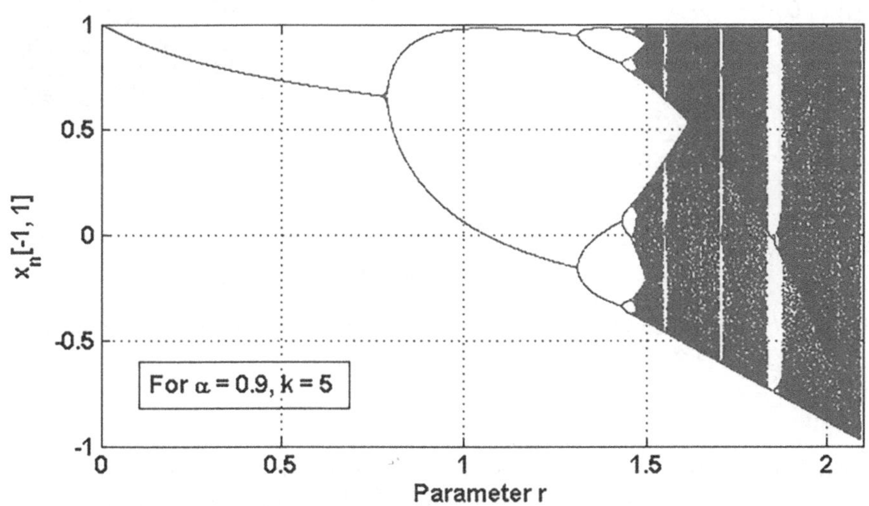

Figure 11.9 Bifurcation plot of the logistic-type map in the modified Mann orbit when $k = 5, \alpha = 0.9$, and $0 \le r \le 2.1$.

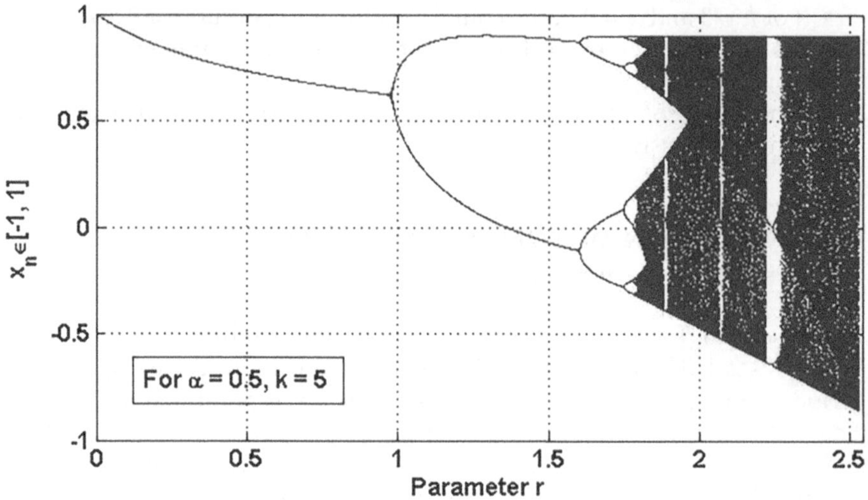

Figure 11.10 Bifurcation plot of the logistic-type map in the modified Mann orbit when $k = 5, \alpha = 0.5$, and $0 \le r \le 2.5$.

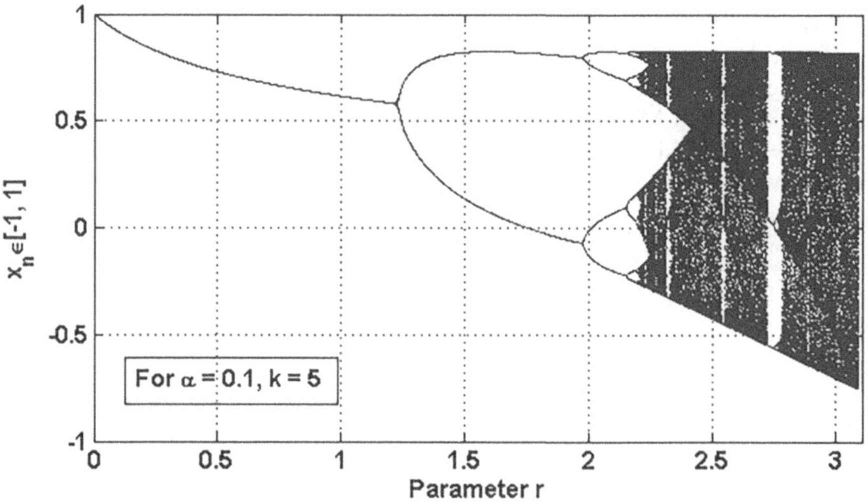

Figure 11.11 Bifurcation plot of the logistic-type map in the modified Mann orbit when $k = 5, \alpha = 0.1$ and $0 \le r \le 3.11$.

growth-rate parameter $0 \le r \le 3.11$ at $k = 5$ and $\alpha = 0.1$. Also, the fixed-point regime occurs at $0 \le r \le 1.23$, which is the largest regime compared to Figures 11.9 and 11.10. Similarly, it shows periodicity and chaotic regime for $1.23 < r \le 2.21$ and $2.21 < r \le 3.11$, respectively.

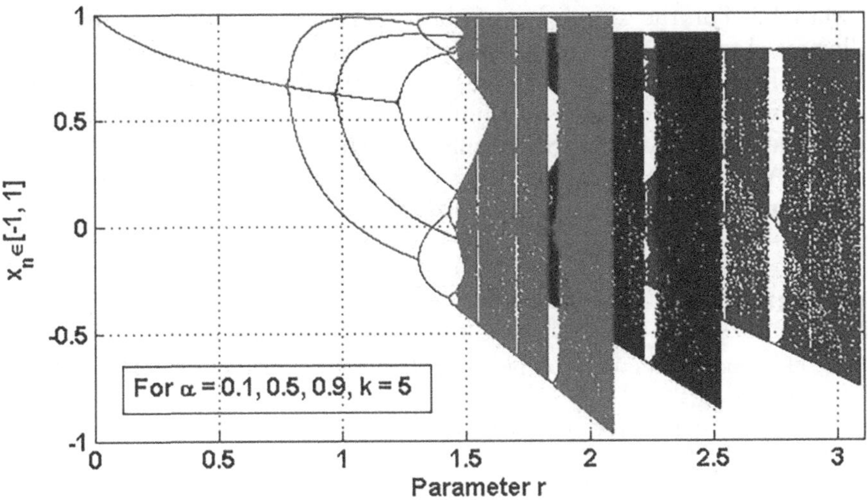

Figure 11.12 Comparative analysis versus bifurcation plots of the logistic-type map in the modified Mann orbit for $k = 5$ and $\alpha = 0.9, 0.5, 0.1$.

From Figures 11.9–11.11 and the comparison in Figure 11.12, the following analysis is carried out for the logistic-type map using the modified Mann orbit:

- For $k = 5$ and $\alpha = 0.1$, the complete bifurcation occurs for $0 \le r \le 3.11$.
- For $k = 5$ and $\alpha = 0.5$, the complete bifurcation occurs for $0 \le r \le 2.5$.
- For $k = 5$ and $\alpha = 0.9$, the complete bifurcation occurs for $0 \le r \le 2.10$.
- As α decreases from 1 to 0, the growth-rate parameter r increases continuously.
- Also, the range of the fixed, periodic, and chaotic states increases simultaneously.
- As compared to earlier Sections 11.4 and 11.5 for $k = 2$ and 3, the overall growth-rate parameter range decreases continuously as the value of the parameter k increases through 5. Therefore, it is noticed that as the range of the growth-rate parameter increases through 2, 3, 5 ..., and so on, the overall growth-rate parameter range decreases with all the dynamical properties like fixed state, periodic state, and chaotic regime.

11.7 CONCLUSION

In this chapter, the dynamical properties of the logistic-type map are examined using a modified Mann orbit framework. The dynamical behavior depends on the three control parameters k, α, and the growth-rate parameter r. The experimental study is carried out using the computer software

"Matlab." For the simplicity of the experimental study, the analysis is illustrated for $k > 1$ and $\alpha = 0.9, 0.5, 0.1$. Further, it is examined that as the value of the parameter k is increased through 1, the range of the growth-rate parameter r is decreased simultaneously. A comparative study versus parameter k, α, and r is also determined, as shown in Figures 11.4, 11.8, and 11.12. As the control parameter k is decreased from 1 to 0, the growth-rate parameter r increases in its prescribed range.

REFERENCES

1. H. Poincare, *Les Methods Nouvells de la Mecanique Leleste*, Gauthier Villars (1899).
2. R. May, Simple mathematical models with very complicated dynamics, *Nature*, (261) (1976), pp. 459–475.
3. E. N. Lorenz, Deterministic nonperiodic flows, *J. Atoms. Sci.*, (20) (1963), pp. 130–141.
4. M. J. Feigenbaum, Quantitative universality for a class of nonlinear transformations, *J. Stat. Phys.*, (19) (1) (1978), pp. 25–52.
5. R. L. Devaney, *An Introduction to Chaotic Dynamical Systems*, 2nd Edition, Addison-Wesley (1948).
6. R. L. Devaney, *A First Course in Chaotic Dynamical Systems: Theory and Experiment*, Addison-Wesley (1992).
7. R. A. Holmgren, *A First Course in Discrete Dynamical Systems*, Springer Verlag, New York Inc. (1994).
8. M. Andrecut, Logistic map as a random number generator, *Int. J. Mod. Phys. B*, (12) (9), (1998), pp. 921–930.
9. R. Chugh, M. Rani and Ashish, Logistic map in Noor orbit, *Chaos Complex. Lett.*, (6) (3) (2012), pp. 167–175.
10. K. T. Alligood, T. D. Sauer and J. A. Yorke, *Chaos: An Introduction to Dynamical Systems*, Springer Verlag, New York Inc. (1996).
11. N. Song and J. Meng, Research on logistic mapping and synchronization, *Proc. IEEE Intell. Contr. Autom.*, (1) (1996), pp. 987–991.
12. N. Singh and A. Sinha, Chaos based secure communication system using logistic map, *Opt. Lasers Eng.*, (48) (2010), pp. 398–404.
13. A. G. Radwan, On some generalized discrete logistic map, *J. Adv. Res.*, (4) (2013), pp. 163–171.
14. B. Parasad and K. Katiyar, A stability analysis of logistic model, *Int. J. Nonlinear Sci.*, (17) (1) (2014), pp. 71–79.
15. W. R. Mann, Mean value methods in iteration, *Proc. Am. Math. Soc.*, (4) (1953), pp. 506–510.
16. Ashish, J. Cao and R. Chugh, Controlling chaos using superior feedback technique with applications in discrete traffic models, *Int. J. Fuzzy Syst.*, (21) (5) (2019), pp. 1467–1479.
17. Ashish, J. Cao and R. Chugh, Chaotic behavior of logistic map in superior orbit and an improved chaos-based traffic control model, *Nonlinear Dyn.*, (94) (2) (2018), pp. 959–975.

18. G. C. Wu, D. Baleanu, Discrete fractional logistic map and its chaos, *Nonlinear Dyn.*, (75) (2014), pp. 283–286.
19. G. C. Wu, D. Baleanu, Discrete chaos in fractional delayed logistic map. *Nonlinear Dyn.*, (80) (2015), pp. 1697–1703.
20. Ashish and J. Cao, A novel fixed-point feedback approaches studying the dynamical behavior of standard logistic map, *Int. J. Bifurcat. Chaos*, (29) (1) (2019), p. 16.
21. Ashish, J. Cao and F. Alsaadi, Chaotic evolution of difference equations in Mann orbit, *J. Appl. Anal. Comput.*, (11) (6) (2021), pp. 1–20.
22. Ashish, J. Cao, F. Alsaadi and A. K. Malik, Discrete superior hyperbolicity in chaotic maps, *Chaos Theory Appl.*, (3) (1) (2021), pp. 34–42.
23. Ashish, J. Cao and R. Chugh, Discrete chaotification in modulated logistic system, *Int. J. Bifurcat. Chaos*, (31) (5) (2021), 2150065 (14 pages).
24. Ashish and Renu Chugh, On the dynamics of a discrete difference map in Mann orbit, *Comput. Appl. Math.*, (41) (226) (2022), pp. 1–19.

Chapter 12

PINN framework for advection-diffusion models

Pooja Rani, Sharanjeet Dhawan, and Vishal Singh

12.1 INTRODUCTION

The advection–diffusion equation plays a central role in describing transport phenomena across physics, engineering, environmental sciences, and biological systems. It governs the combined effects of advection, which represents the directed movement of a substance due to bulk flow, and diffusion, which models spreading due to molecular interactions or random mixing. Classical applications include transport of heat, solute dispersion in groundwater, atmospheric pollutant migration, contaminant spreading in rivers, and density evolution in traffic flow systems. Because these processes exhibit both spatial and temporal variability, obtaining accurate and computationally efficient solutions to the advection–diffusion equation is crucial for prediction, control, and inverse analysis in real-world systems.

The advection-diffusion equation plays a vital role in modeling transport phenomena, yet solving it remains a challenge, especially for complex real-world systems. Traditionally, it has been approximated using conventional numerical methods, including the finite-difference method (FDM), finite volume method (FVM), and finite element method (FEM). Although these methods are mathematically robust, they require fine spatial discretization and may suffer from artificial diffusion, numerical oscillations, and instability in advection-dominated regimes. Their performance also degrades when data are sparse or noisy, particularly in inverse problems where unknown physical parameters such as velocity, diffusivity, or source strength must be estimated. Moreover, mesh generation becomes increasingly complex for higher-dimensional and irregular domains, leading to high computational costs.

The development of Physics-Informed Neural Networks (PINNs) has presented a viable mesh-free partial differential equation solution option in recent years. PINNs,introduced by Raissi et al. (2019), calculate derivatives effortlessly and incorporate governing equations, boundary conditions, and

 DOI: 10.1201/9781003521389-12

initial conditions during training using automatic differentiation. This eliminates the need for big labeled datasets and allows the network to discover solutions that naturally satisfy physical restrictions. Strong generalization features and data efficiency have been demonstrated by PINNs' successful applications in fluid mechanics (Raissi et al., 2019), which demonstrated the ability to reconstruct velocity and pressure fields from flow visualization, heat transfer, anomalous diffusion, stochastic partial differential equations (PDEs), and inverse modeling.

In the context of transport processes, PINNs have shown significant potential for modeling convection-diffusion systems, pollutant migration in porous media, subsurface flow, and thermo-fluid interactions. Studies reveal that PINNs can outperform traditional numerical solvers in scenarios involving incomplete data, unknown parameters, multi-scale behavior, or irregular computational domains. However, challenges remain related to stiff advection-dominated dynamics, loss imbalance, and training instability, motivating further research into reliable PINN-based solvers. This work directly addresses these challenges by incorporating Fourier feature embedding to resolve sharp boundary layers, residual-based adaptive refinement (RAR) to dynamically concentrate collocation points in high-error regions, physically motivated loss weights to enforce boundary and initial conditions strongly, and causal training weights to ensure correct temporal propagation over long time horizons.

In this work, we employ a mesh-free PINN framework to obtain forward and inverse solutions of the 1D advection–diffusion equation under physically relevant scenarios. Two case studies are considered:

1. Atmospheric pollutant dispersion from a continuous source.
2. Traffic flow dynamics, where vehicle density evolution is modeled as a convection–diffusion process.

In both cases, the PINN is trained to satisfy the governing PDEs, initial conditions, and boundary conditions simultaneously. The resulting predictions are compared with analytical and numerical benchmark solutions using error norms, pointwise comparisons, and 3D surface visualizations. The results demonstrate that the PINN model accurately reconstructs the underlying transport fields and is effective for parameter inference in inverse settings.

This chapter highlights the strengths of PINNs as a flexible and data-efficient computational tool, particularly for advection-dominated and data-scarce problems, where traditional mesh-based methods struggle. The suggested method offers a reliable and accurate framework for simulating a variety of transport phenomena in environmental, engineering, and traffic systems by directly incorporating physical rules into neural architectures.

12.2 METHODOLOGY

A PINN framework serves as the foundation for solving the advection-diffusion equation. This method allows the neural network to approximate the solution while precisely satisfying the underlying PDEs and the specified initial and boundary conditions, since the governing physical laws are directly included in the learning process. The general PDE is:

$$\frac{\partial u}{\partial t} + a\frac{\partial u}{\partial x} = D\frac{\partial^2 u}{\partial x^2}$$

where $u(x,t)$ is the transported quantity, a is the advection velocity, and D is the diffusion coefficient. The equation is solved over a spatial domain $x \in \Omega \subset R$ and a temporal interval $t \in [0,T]$.

12.2.1 Physics-informed neural network architecture

To approximate $u(x,t)$, a fully connected neural network with multiple hidden layers is employed. The network produces u after receiving (x,t) as the input. Two neurons that represent the spatial coordinate x and time t make up the input layer. These inputs are routed through several hidden layers, each of which has a set number of neurons with nonlinear activation functions. The hyperbolic tangent is chosen as the activation function for these hidden layers due to its smooth, infinitely differentiable nature, which is a critical requirement for PINNs since automatic differentiation computes high-order derivatives of the network output to form the PDE residuals. Unlike the Relu, which is piecewise linear and has discontinuous second derivatives, making it unsuitable for second-order terms like diffusion, tanh provides continuous derivatives across all orders, ensuring stable gradient flow and a more accurate approximation of differential operators. The scalar prediction is generated by a single neuron in the output layer that has a linear activation function. The necessary temporal and spatial derivatives of the network output are then calculated using automatic differentiation. To help the network capture sharp gradients and high-frequency spatial features, a Fourier feature embedding layer is added before the first hidden layer. The raw inputs (x,t) are transformed as $\gamma(z) = [\cos(Bz), \sin(Bz)]$, where B is a fixed random matrix sampled from $N(0,\sigma 2)$ with $m = 64$ features. This overcomes the tendency of standard networks to learn only smooth, low-frequency features, which is a known limitation when solving PDEs with sharp gradients or thin boundary layers (Wang et al., 2021).

12.2.2 Loss function formulation

The total loss function is defined as a weighted sum of multiple components:

$$L(\theta) = \omega_{PDE}L_{PDE} + \omega_{IC}L_{IC} + \omega_{BC}L_{BC}$$

where L_{PDE}, L_{IC}, and L_{BC} represent the losses associated with the governing equation, initial conditions, and boundary conditions, respectively. The coefficients ω_{PDE}, ω_{IC}, and ω_{BC} are weighting parameters that balance the contributions of each term. In this work, the weights are chosen based on the physical difficulty of each problem rather than using equal weights. Elevated weights on the boundary and initial condition terms are found to be essential in advection-dominated problems, where weak enforcement of boundary constraints leads to unphysical predictions in the interior of the domain. To assess robustness, different initial conditions are considered, including Gaussian distributions, step functions, and sinusoidal profiles, each representing physically relevant transport scenarios. Furthermore, depending on the problem formulation, different boundary conditions such as Dirichlet, Neumann, or periodic conditions are imposed.

12.2.3 Training procedure

The goal of training is to minimize the overall loss function with respect to the network parameters θ. To guarantee both stability and convergence, a two-stage optimization approach is used. The Adam optimizer is used in the first step because of its robustness and capacity to effectively explore the parameter space during early training. This phase aids the network in quickly minimizing the loss and locating an appropriate area within the optimization landscape. By utilizing second-order information, the L-BFGS optimizer, a quasi-Newton technique, is employed in the second stage to increase accuracy and speed of convergence. Collocation points are equally sampled in both space and time, and extra adaptive sampling is used close to areas with sharp fronts or steep gradients, including advection-dominated layers. Beyond fixed clustering, a RAR strategy is also applied dynamically during training. Every few thousand epochs, the PDE residual is evaluated at 10,000 candidate points, and the 500 points with the largest residual are permanently added to the collocation pool, automatically focusing training effort in regions where the physics is hardest to satisfy (Lu et al., 2021). Additionally, a cosine annealing schedule decays the learning rate smoothly from $\eta_{max} = 10^{-3}$ to $\eta_{min} = 10^{-5}$, avoiding the oscillations observed with a fixed learning rate. The two-phase training first uses Adam for global exploration and then L-BFGS with Strong Wolfe line search for tight local convergence (Raissi et al., 2019).

This targeted sampling strategy improves the network's ability to resolve localized features without significantly increasing the overall computational cost. The training process continues until convergence criteria, such as loss reduction or a maximum iteration limit, are satisfied. The individual loss components are tracked during training to assess how effectively the enforcement of the governing PDE and the satisfaction of boundary and initial constraints are balanced.

12.2.4 Verification and benchmarking

The analysis proceeds by considering two cases for verification and benchmarking. The proposed solutions are evaluated against analytical formulas and finite-difference benchmarks, with performance measured via mean-squared error (MSE) and root mean-squared error (RMSE).

12.2.4.1 Case 1: pollutant plume

The evolution of a contaminant concentration $C(x,t)$ over time and space under the combined effects of atmospheric mixing and wind-driven transport is described by a one-dimensional transient advection-diffusion problem representing pollutant transport in an atmospheric channel of length L. The pollutant enters the system at the left channel boundary and travels rightward at a constant advection velocity c while also spreading due to diffusion, which is indicated by the diffusion coefficient D. The concentration gradually decreases toward zero near the outlet, as the particles leave the system or become diluted while moving through the channel. The governing equation describing this behavior is given by:

$$\frac{\partial C}{\partial t} + c\frac{\partial C}{\partial x} = D\frac{\partial^2 C}{\partial x^2}, x \in [0,L], t \in [0,T]$$

where $C(x,t)$ represents the contaminant concentration, c represents the advection velocity, and D represents the diffusion coefficient.
Parameters: $c = 0.1, D = 0.01, L = 10, T = 100$
Initial conditions: $C(x,0) = 0$
Boundary conditions:

$$u(0,t) = C_{Source} = 1 \text{ (constant pollutant source at inlet)}$$
$$u(L,t) = 0 \text{ (far-field zero-concentration condition)}$$

12.2.4.2 Case 2: traffic flow

In this case, a one-dimensional transient traffic flow model is considered, where vehicle density changes over time and space, controlled by an advection-diffusion-type equation. The average vehicle speed is represented by the advection term, while driver interaction and density smoothing effects are taken into account by the diffusion term. Road endpoints are subject to Neumann boundary conditions, whereas the initial condition establishes the initial vehicle density distribution along the road. To assess accuracy, the predicted solution is contrasted with an analytical reference solution based on a Gaussian profile. The reference solution is the exact finite-domain solution obtained using the method of images to satisfy the Neumann boundary conditions. Mirror sources are placed at $\pm x_0 + 2nL \, for \, n = -25,\ldots,25$, giving

$$u(x,t) = \frac{\sigma_0}{\sqrt{\sigma_t^2}} \sum_{n=-25}^{25} \left[\exp\left(-\frac{(x-\mu_1^n)^2}{2\sigma_t^2} \right) + \exp\left(-\frac{(x-\mu_2^n)^2}{2\sigma_t^2} \right) \right]$$

where $\mu_1^n = x_0 + 2nL + vt$, $\mu_2^n = -x_0 + 2nL + vt$, and $\sigma_t^2 = \sigma_0^2 + 2Dt$ (Haberman, 2004). This formulation correctly accounts for boundary reflections that develop at later times $t > 10$, which are not captured by the infinite-domain Gaussian solution.

One-dimensional advection-diffusion equation for vehicle density $u(x,t)$ with mean traffic velocity v and diffusion coefficient D is represented as:

$$\frac{\partial u}{\partial t} + v\frac{\partial u}{\partial x} = D\frac{\partial^2 u}{\partial x^2}, x \in [0,L], t \in [0,T]$$

where $u(x,t)$ represents the vehicle density, v represents the mean traffic velocity, and D represents the diffusion coefficient. We take $v = 0.2, D = 0.05,$ $L = 10,\ T = 20$ with $u(x,0) = e^{\frac{-1}{2}\left(\frac{x-x_0}{\sigma}\right)^2}$ $u_x(0,t) = 0,\ u_x(L,t) = 0,$ where $x_0 = \frac{L}{2} = 5, \sigma = 0.5$. The solution $u(x,t)$ is approximated using a fully con nected neural network $u_\theta : R^2 \to R$, where θ represents the trainable network parameters. The neural network architecture used for the pollutant plume case consists of a fully connected network with the layer configuration [2,100,100,100,100,1]. For the traffic flow case, a Fourier feature embedding layer is first applied to the inputs, followed by a fully connected network with the configuration [2,128,128,128,128,128,1]. The Fourier embeddings use scaling parameters of $\sigma = 5.0$ and $\sigma = 3.0$ for the pollutant plume and traffic flow cases, respectively. In both cases, the hyperbolic tangent (tanh) activation function is used for the hidden layers, and Xavier initialization is applied to all network layers. For improved performance, x and t are normalized to $[-1,1]$ or $[0,1]$ before passing them to the networks. The network prediction is denoted as $u_\theta(x,t)$. For both cases, the PDE residual is:

$$R(x,t;\theta) = u_{\theta,t} + a u_{\theta,x} - D u_{\theta,\,xx}$$

where a represents c (wind speed) in the pollutant plume case or v (traffic velocity) in the traffic flow case. The required derivatives are then computed using automatic differentiation.

Total loss is a weighted sum:

$$L(\theta) = \omega_{PDE} L_{PDE} + \omega_{IC} L_{IC} + \omega_{BC} L_{BC}$$

where the coefficients ω_{PDE}, ω_{IC}, and ω_{BC} are weighting parameters that balance the contributions of the PDE residual, initial condition, and boundary-condition losses, respectively. In this work, the loss weights are set based on the physical difficulty of each case. For the pollutant plume case, $w_{PDE} = 1$, $w_{IC} = 50$, and $w_{BC} = 500$. The large boundary-condition weight is necessary at $Pe = 100$ to prevent the network from violating the Dirichlet constraints at the inlet and outlet. For the traffic flow case, $w_{PDE} = 1$, $w_{IC} = 50$, and $w_{BC} = 100$.

The PDE loss is the mean-squared residual at N_f collocation points:

$$L_{PDE} = \frac{1}{N_f} \sum_{i=1}^{N_f} | R(x_i^{(f)}, t_i^{(f)}; \theta |^2$$

For the pollutant plum case, the initial condition loss is given as follows:

$$L_{IC} = \frac{1}{N_{ic}} \sum_{i=1}^{N_{ic}} \left| u_\theta(x_i, t_{ic}) - 0 \right|^2$$

where $t_{ic} = 1 \times 10^{-4}$ is used to avoid singular behavior at $t = 0$.

For the traffic flow case, the initial condition loss is given as follows:

$$L_{IC} = \frac{1}{N_{ic}} \sum_{i=1}^{N_{ic}} \left| u_\theta(x_i, 0) - e^{-\frac{1}{2}\left(\frac{x - x_0}{\sigma}\right)^2} \right|^2$$

The boundary-condition loss is defined as follows:

$$L_{BC} = \frac{1}{N_{bc}} \sum_{i=1}^{N_{bc}} \left| B\left(u_\theta(x_{bc}, t_{bc})\right) \right|^2$$

where B stands for the boundary operator, which enforces zero gradient under Neumann and specified values under Dirichlet conditions.

Model training is performed using the Adam optimizer with a learning rate of 1×10^{-3}. For the pollutant plume case, 5,000 collocation points are used per epoch, with half clustered near $x = L$ within the boundary layer of thickness $\delta = \max(5D / c, 0.5)$, and training runs for 20,000 Adam epochs, followed by 3,000 L-BFGS iterations. For the traffic flow case, collocation points are split into three equal groups, uniform over the full domain, clustered near the initial Gaussian support at $x_0 \pm 2\,\sigma_0$ and concentrated at late times $t \in [T/2, T]$, and training runs for 30,000 Adam epochs, followed by 5,000 L-BFGS iterations. For the traffic flow case, a causal weighting term $w(t) = \exp(-\varepsilon \cdot t / T)$ with $\varepsilon = 5.0$ is applied to the PDE loss, assigning higher weight to early times so that the network learns the solution sequentially from $t = 0$ forward (Wang et al., 2022). All loss components are tracked independently in order to monitor training progress and convergence behavior. MSE and RMSE are computed between the PINN prediction and the analytic/approximate solution on a dense grid:

$$MSE = \frac{1}{N}\sum_{i=1}^{N}\left|u_\theta\left(x_i, t_{ic}\right) - u_{ref}\left(x_i, t_{ic}\right)\right|^2, RMSE = \sqrt{MSE}$$

The relative L_2 error is also computed:

$$Rel\, L_2 = \frac{\left\|u_\theta - u_{ref}\right\|^2}{\left\|u_{ref}\right\|^2}$$

Additionally, line plots at selected time slices and 3D surface plots of the solution versus x and t are visualized for comparison.

12.3 RESULTS

For the pollutant plume case, the semilogarithmic plot of the training loss shows a rapid decrease during the early stages of optimization in Figure 12.1. The loss initially starts at approximately 10^3 and decreases to around $10^{-1} - 10^{-2}$ during the Adam optimization phase. While the Adam phase exhibits noticeable fluctuations, it steadily drives the network toward a lower-loss region. At iteration 50,000, the optimization switches from Adam to L-BFGS, which leads to a sharp and smooth reduction in the loss. During this second phase, the loss quickly drops by several additional orders ofs magnitude, reaching values on the order of 10^4. This two-stage optimization strategy produces a characteristic loss profile, with gradual stochastic descent during the Adam phase followed by rapid deterministic convergence during the L-BFGS phase, clearly visible at the transition point in the plot.

Semilogarithmic plot of the training loss for the traffic flow case showing the two-stage optimization strategy using Adam followed by L-BFGS can be seen through Figure 12.2. During the Adam phase (iterations $0 - 30{,}000$), the loss decreases rapidly from approximately 10^2 to around 10^{-5}, with noticeable oscillations throughout training. These oscillations arise from stochastic optimization and the periodic addition of new high-residual collocation points through the RAR strategy. Despite these spikes, the overall downward trend indicates stable convergence as the network progressively satisfies the PDE, initial condition, and boundary-condition constraints. At iteration 30,000, the optimizer switches to L-BFGS, which produces a further smooth reduction in the loss, driving it below 10^{-5}. The clear transition between the stochastic Adam phase and the deterministic L-BFGS refinement phase highlights the effectiveness of the hybrid optimization strategy in achieving stable and accurate PINN convergence.

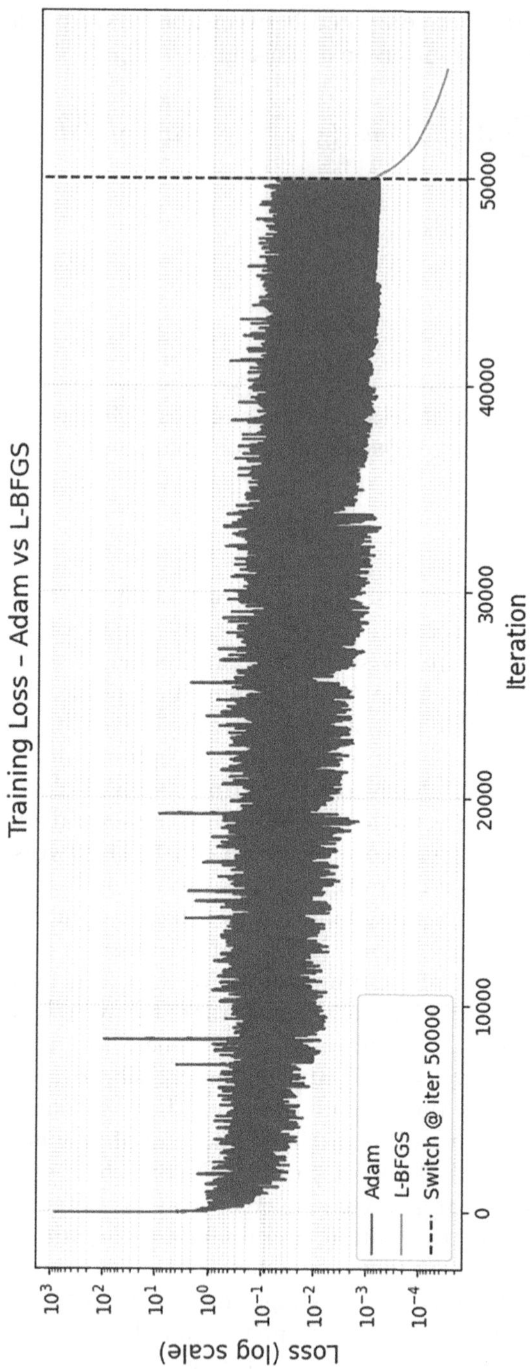

Figure 12.1 Semi-logarithmic plot of the training loss for pollutant plume case.

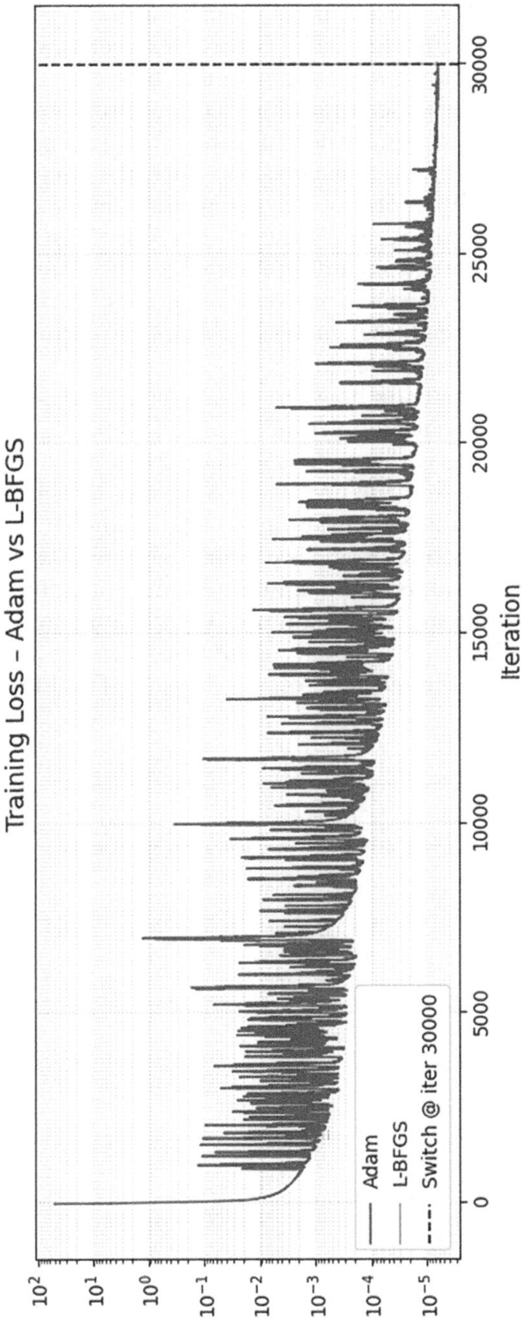

Figure 12.2 Semi-logarithmic plot of the training loss for traffic flow case.

Comparison of the analytical steady-state solution and the PINN prediction at $t = T = 100$ for the pollutant plume case across the full spatial domain is shown in Figure 12.3. The two curves overlap closely over most of the domain, indicating excellent agreement between the PINN and the analytical solution. The concentration remains approximately constant for most of the domain and rapidly decreases near the outlet at $x = 10$, consistent with the imposed zero-concentration Dirichlet boundary condition. The PINN successfully captures the sharp gradient associated with the diffusive boundary layer near the outlet, demonstrating that the enhanced training strategy, including Fourier feature embedding, a deeper network architecture, clustered collocation points, RAR, and increased boundary-condition weighting, enables accurate resolution of the underlying advection–diffusion physics.

In Figure 12.4, comparison between the analytical finite-domain solution and the PINN predictions at multiple time instances $(t = 0.1, 5, 10, 15, 18, \text{and } 20)$ for the traffic flow case is showcased. At the early time $(t = 0.1)$, the PINN prediction closely matches the analytical solution, accurately reproducing the initial Gaussian profile. As time progresses $(t = 5 \text{ and } t = 10)$, the advected and diffusing concentration profile is captured very well by the PINN, with strong agreement in peak position, amplitude, and spreading. At later times $(t = 15, 18, \text{and } 20)$, the analytical solution shows boundary reflection near $x = 0$ due to the Neumann boundary condition, forming a secondary concentration profile on the left side of the domain. The PINN successfully

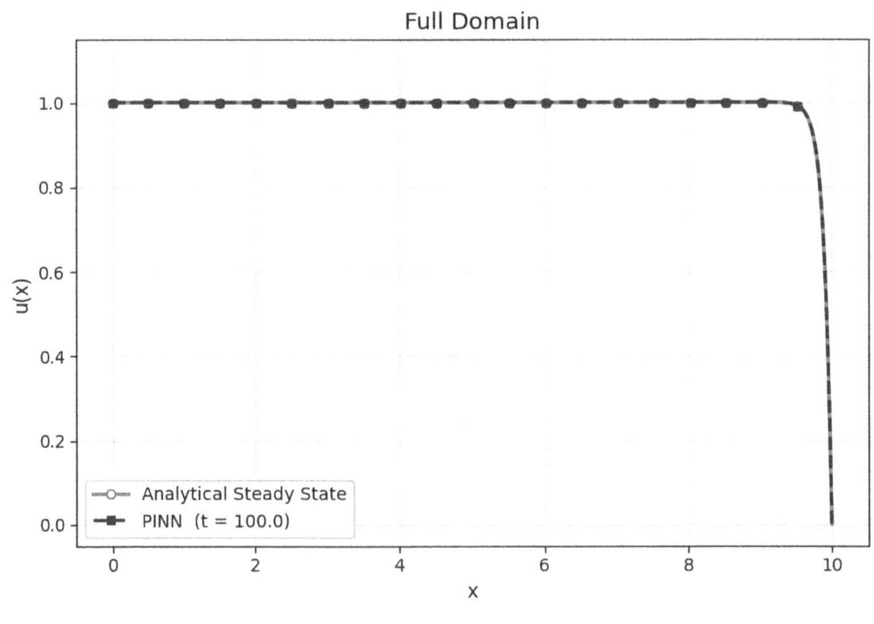

Figure 12.3 Analytical solution vs. PINN prediction for pollutant plume case at t = 100.

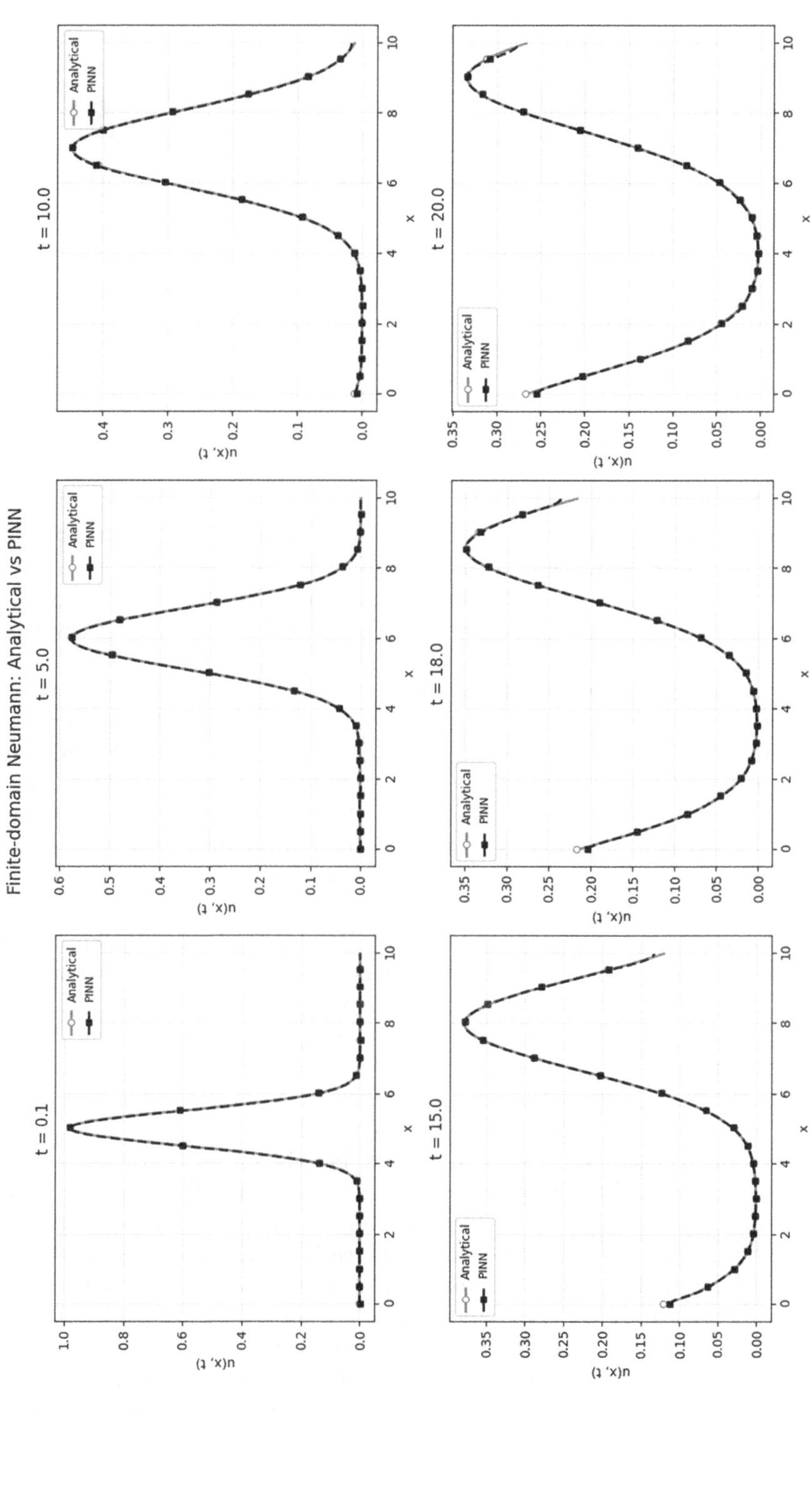

Figure 12.4 Analytical solution vs. PINN prediction for traffic flow case at t = 0.1, 5, 10, 15, 18, 20.

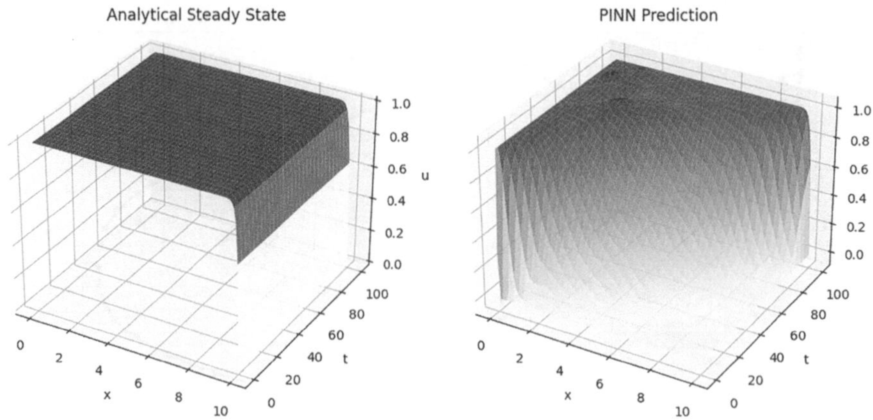

Figure 12.5 3D comparison of Analytical solution and PINN solution for pollutant plume case.

reproduces both the primary advected peak and the reflected component, maintaining close agreement with the analytical solution across the entire domain. Overall, the results demonstrate that the PINN accurately captures the advection–diffusion dynamics and the boundary reflection behavior over time.

Comparison of the analytical steady-state surface and the PINN prediction for the pollutant plume case over the full space–time domain can be seen through Figure 12.5. The analytical solution shows a nearly uniform concentration close to unity across most of the domain, with a sharp drop near the outlet $(x \approx 10)$ due to the imposed Dirichlet boundary condition. The PINN prediction reproduces the overall spatiotemporal behavior and the downstream propagation of concentration from the inlet while also capturing the formation of the boundary layer near the outlet. The surface evolution indicates that the network correctly learns the transient progression toward the steady-state profile. Minor deviations appear in the sharp gradient region near the outlet and in the early transient region, where the PINN surface appears slightly smoother than the analytical solution. Overall, the results demonstrate that the PINN successfully captures the dominant advection–diffusion dynamics and the transition from the initial condition to the steady-state concentration profile across the domain.

Three-dimensional surface comparison between the analytical finite-domain solution and the PINN prediction for the traffic flow case over the full space–time domain can be seen in Figure 12.6. The analytical surface shows the Gaussian pulse advecting to the right while gradually spreading due to diffusion. The PINN prediction closely reproduces this space–time evolution, accurately capturing both the translation of the peak and the increasing spread of the profile over time. The overall shape and trajectory of the pulse match well between the two surfaces, indicating that the PINN successfully learns the coupled advection–diffusion dynamics without

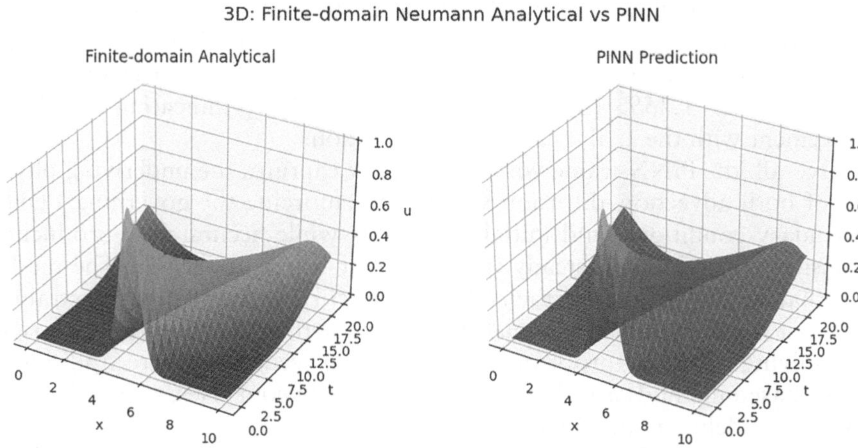

Figure 12.6 3D comparison of Analytical solution and PINN solution for traffic flow case.

introducing numerical instability. Minor smoothing is visible in the PINN surface, but the dominant transport behavior and temporal evolution are well preserved.

The resulting error metrics indicate good quantitative agreement between the PINN predictions and the analytical reference solutions. For the pollutant plume case, the obtained values are $RMSE = 6.387 \times 10^{-1}$ and $MSE = 4.079 \times 10^{-1}$, while for the traffic flow case, the errors are significantly smaller, with $RMSE = 1.459 \times 10^{-3}$ and $MSE = 2.130 \times 10^{-6}$. These values are relatively small compared to the magnitude of the solution, demonstrating that the PINN accurately approximates the underlying advection–diffusion dynamics. The remaining discrepancies are primarily attributed to boundary-domain effects and small diffusion-related variations rather than limitations of the model itself. Overall, the results confirm that the proposed PINN framework provides a reliable and physically consistent approximation of the governing transport processes.

12.4 CONCLUSION

This study presents a PINN framework for solving the one-dimensional advection–diffusion equation, demonstrated through two case studies: atmospheric pollutant dispersion and traffic flow dynamics. In the pollutant plume case, the PINN exhibited stable convergence, with the training loss decreasing from approximately 10^1 to 10^{-3} over 20,000 epochs. The model successfully captured the inlet source concentration and enforced the zero-concentration condition at the outlet. Quantitative evaluation yielded $RMSE = 6.387 \times 10^{-1}$ and $MSE = 4.079 \times 10^{-1}$, with the primary discrepancies occurring in the downstream diffusive boundary layer region $(x \approx 7 - 9)$.

For the traffic flow case, the PINN accurately captured both the advection-driven pulse propagation and the diffusion-induced spreading of the concentration profile. The model achieved significantly lower error values, with $RMSE = 1.459 \times 10^{-3}$ and $MSE = 2.130 \times 10^{-6}$, indicating excellent agreement with the analytical reference solution.

Overall, the PINN framework successfully captures the underlying physics of both advection–diffusion systems by enforcing the governing PDEs, boundary conditions, and initial conditions while accurately reproducing transient evolution and steady-state behavior where applicable. The mesh-free nature of PINNs and their ability to incorporate physical constraints without requiring labeled simulation data make them a promising approach for solving transport problems, particularly in scenarios where traditional mesh-based methods are challenged by complex geometries, sparse data, or inverse modeling requirements.

Several methodological enhancements were incorporated in this work to improve model accuracy and stability. Fourier feature embedding mitigates spectral bias and enables the network to resolve high-frequency features in the high Péclet number pollutant plume case $(Pe = 100)$. RAR dynamically concentrates collocation points in regions with large residuals during training. Physically motivated loss weights ensure strong enforcement of boundary and initial conditions. A two-stage optimization strategy combining Adam and L-BFGS, together with cosine annealing, provides stable and efficient convergence. Additionally, causal training weights enable accurate temporal propagation in the traffic flow case over the long time horizon $(T = 20)$. The method of images is used to construct a physically consistent finite-domain analytical reference solution for the Neumann boundary-condition case. These combined improvements result in reduced error metrics and demonstrate the effectiveness of the PINN framework for modeling advection–diffusion transport phenomena.

BIBLIOGRAPHY

Cybenko, G. (1989). Approximation by superpositions of a sigmoidal function. *Mathematics of Control, Signals, and Systems*, 2(4), 303–314. https://doi.org/10.1007/BF02551274

Dissanayake, M. G., & Phan-Thien, N. (1994). Neural-network-based approximations for solving partial differential equations. *Communications in Numerical Methods in Engineering*, 10(3), 195–201. https://doi.org/10.1002/cnm.1640100303

Goodfellow, I., Bengio, Y., & Courville, A. (2016). *Deep learning*. MIT Press. https://doi.org/10.1007/BF02551274

Haberman, R. (2004). *Applied partial differential equations with Fourier series and boundary value problems*. Pearson.

Karniadakis, G. E., Kevrekidis, I. G., Lu, L., Perdikaris, P., Wang, S., & Yang, L. (2021). Physics-informed machine learning. *Nature Reviews Physics*, 3(6), 422–440. https://doi.org/10.1038/s42254-021-00314-5

Kelley, C. T. (1995). *Iterative methods for linear and nonlinear equations*. Society for Industrial and Applied Mathematics. https://doi.org/10.1137/1.9781611970944

Lagaris, I. E., Likas, A., & Fotiadis, D. I. (1998). Artificial neural networks for solving ordinary and partial differential equations. *IEEE Transactions on Neural Networks*, 9(5), 987–1000. https://doi.org/10.1109/72.712178

LeVeque, R. J. (2007). Finite difference methods for ordinary and partial differential equations: steady-state and time-dependent problems. *Society for Industrial and Applied Mathematics*. https://doi.org/10.1137/1.9780898717839

Lu, L., Meng, X., Mao, Z., & Karniadakis, G. E. (2021).DeepXDE: A deep learning library for solving differential equations. *SIAM Review*, 63(1), 208–228. https://doi.org/10.1137/19M1274067

Mao, Z., Jagtap, A. D., & Karniadakis, G. E. (2020). Physics-informed neural networks for high-speed flows. *Computer Methods in Applied Mechanics and Engineering*, 360, 112789. https://doi.org/10.1016/j.cma.2019.112789

Morton, K. W., & Mayers, D. F. (1994). *Numerical solution of partial differential equations* (p. 512). Cambridge University. https://doi.org/10.1017/CBO9780511812248

Paszke, A., Gross, S., Massa, F., Lerer, A., Bradbury, J., Chanan, G., ... Chintala, S. (2019). Pytorch: An imperative style, high-performance deep learning library. *Advances in Neural Information Processing Systems*, 32. https://doi.org/10.48550/arXiv.1912.01703

Raissi, M., Perdikaris, P., & Karniadakis, G. E. (2019). Physics-informed neural networks: A deep learning framework for solving forward and inverse problems involving nonlinear partial differential equations. *Journal of Computational Physics*, 378, 686–707. https://doi.org/10.1016/j.jcp.2018.10.045

Stull, R. B. (1988). Mean boundary layer characteristics. In *An introduction to boundary layer meteorology* (pp. 1–27). Dordrecht: Springer Netherlands. https://doi.org/10.1007/978-94-009-3027-8

Tartakovsky, A. M., Marrero, C. O., Perdikaris, P., Tartakovsky, G. D., & Barajas-Solano, D. (2020). Physics-informed deep neural networks for learning parameters and constitutive relationships in subsurface flow problems. *Water Resources Research*, 56(5), e2019WR026731. https://doi.org/10.1029/2019WR026731

Wang, S., Sankaran, S., & Perdikaris, P. (2022). Respecting causality is all you need for training physics-informed neural networks. *arXiv preprint arXiv:2203.07404*. https://doi.org/10.48550/arXiv.2203.07404

Wang, S., Wang, H., & Perdikaris, P. (2021). On the eigenvector bias of Fourier feature networks: From regression to solving multi-scale PDEs with physics-informed neural networks. *Computer Methods in Applied Mechanics and Engineering*, 384, 113938. https://doi.org/10.1016/j.cma.2021.113938

Chapter 13

Content marketing performance measurement through OWA operator

Martín Huesca-Gastélum, Alicia Delgadillo-Aguirre, Luis Pérez-Arellano, Claudia Tirado-Gálvez, and Ernesto León-Castro

13.1 INTRODUCTION

The growing interaction between the consumer and the company has caused private sector organizations around the world to redesign their strategies and adapt to new roles or mechanisms that involve the generation of content (Hollebeek & Macky, 2019). As Baltes (2015) emphasizes, to remain competitive, these institutions had to make content marketing their primary marketing approach. Likewise, this new framework has been positioned as an essential element within digital marketing because it allows companies to adjust their communication to reach their customers more effectively (Du Plessis, 2017; Taiminen & Ranaweera, 2019).

These changes have forced marketing executives to become more market-oriented in their decision-making, which requires formal means to acquire accurate and timely information (Bala & Verma, 2018). Hence, it has become so popular and has received increasing interest from academics and marketers for its ability to attract customers through attractive and valuable information for their needs (Bangsawan et al., 2021), but with this also arises the need to create, distinguish, and share relevant content to attract more customers (Taiminen & Ranaweera, 2019).

Therefore, it is important to develop constructs that quantify and analyze their contribution to companies' business objectives. However, the existing literature on its measurement is still scarce, since there are very few empirical studies that consider metrics to measure their impact on company performance (Koob, 2021; Müller & Christandl, 2019; Rancati & Gordini, 2014). In other words, as Hollebeek & Macky (2019) comment, despite the significant interest of professionals and managers, academic research on its measurement is lagging, which generates a knowledge gap regarding this issue.

Furthermore, there is no consensus on the definition of content marketing nor regarding the metrics and techniques to be used, due to the presence of different perspectives and methodological practices regarding what exactly constitutes content marketing (Ho et al., 2020). Consequently, to solve this

 DOI: 10.1201/9781003521389-13

limitation and reduce this gap in the existing literature, the objective of this study is to measure the content marketing performance of a company that markets mobile phone accessories through the ordered weighted average operator (OWA) and its extensions. This proposal presents a more flexible and adaptable way of measuring this variable through this fuzzy logic technique.

Furthermore, using aggregation operators allowed not only to classify personalized campaigns based on their content capabilities but also to highlight the relative importance of the different criteria uses to evaluate a marketing campaign. In contrast, the approach proposed by Rancati & Gordini (2014) was used because it is the one that considers and measures the marketing campaign based on different indicators.

Finally, the structure of the paper is the following. In Section 13.2, a review of literature on content marketing performance measurement is presented. Section 13.3 presents the application to a specific case of a company marketing campaign, and Section 13.4 summarizes the main conclusions.

13.2 A REVIEW OF THE LITERATURE ON CONTENT MARKETING PERFORMANCE MEASUREMENT

Due to the Internet, new types of marketing have been presented as a different alternative for doing business. An example of these approaches is content marketing, a tool that has attracted attention for its ability to attract customers through value creation (Mansour & Barandas, 2017). As Taheri & Bashekouh Ajirlo (2023) emphasize, this paradigm shift has positioned online sales and content as a dominant force in business, surpassing traditional techniques.

Despite this growing momentum that has gained worldwide attention, as well as the importance it has acquired, Koob (2021) maintains that this modern framing has been little explored; therefore, there is little empirical evidence not only about its effectiveness, but also with respect to the metrics necessary to measure its impact on the performance of companies. In the words of Wall & Spinuzzi (2018), recent years' research has focused on definitions or conceptualizations and only a small number of analyses have focused on measuring their performance.

One of the most recent studies is the one presented by Ho et al. (2020), who developed a framework to identify the most attractive practices in value creation. To achieve their purpose, the authors analyzed the strategies of five Singapore companies through case studies, interviews, and content analysis. According to the findings, for campaigns to have a positive impact, they must be aligned with the brand's values and oriented to audiences who are willing to consume and share said content.

Similarly, Koob (2021) conceptualizes content marketing to empirically investigate those elements that determine the effectiveness of content strategies. To do so, he collects data from 263 organizations from different sectors and analyzes them through multiple regression. The results indicate that production, as well as the regulations and needs of the target group, has the greatest impact on the effectiveness of content marketing.

On the other hand, Naseri & Noruzi (2018) designed and validated a model that can be used as a guide to measure content marketing performance in different companies and industries. The approach used was a combination of two stages. In the first, they used the meta-synthesis method to identify the dimensions. While, in the second, the Fuzzy Delphi technique was chosen. The results of that project highlight the ability of this framework to establish and implement content marketing strategies in different companies and industries.

Rancati & Gordini (2014) did something similar by laying out the most popular metrics to determine the performance impact of some content creation strategies. The authors began their methodology by reviewing the existing literature on the variable in question, as well as the main metrics used for its measurement. Subsequently, they analyzed the degree of use and effectiveness of the strategy and indicators in a sample of 235 Italian companies, through an integrated questionnaire of 20 items. Its results show the main criteria for measuring the performance of content marketing but also demonstrate the application of these reagents to determine the effectiveness of companies' content strategies.

In contrast, Müller & Christandl (2019) examine how different types of content are perceived and the influence they have on the persuasion of the target audience. The authors developed a serial mediation model applied to users within the video game industry. The findings suggest that user-generated content is perceived positively and has a greater influence on their decisions, while paid or sponsored content is not accepted by customers because they lack credibility and, therefore, is the lowest level of persuasion they achieve.

Another relevant study is the one presented by Naseri et al. (2023) since the authors propose a fuzzy technique to classify the components of content generation. To achieve this, they carried out a literature review to identify the most used criteria and evaluate them through expert judging, as well as the application of the Delphi technique; subsequently, they used the TOPSIS tool to make the final classification of these items. Among the results, the fact that 42 basic components have been identified for evaluating the most effective content for companies stands out because they increase content consumption but also encourage audience participation.

Taheri & Bashekouh Ajirlo (2023) present a similar study by using fuzzy logic to analyze the content strategies of 30 television brands in Iran. The authors collected the opinions of 550 users and analyzed them through the fuzzy inference system (FIS) and the Mamdani inference kernel. Its evaluation highlights Samsung as the television brand with the highest level of performance in its content campaigns. In contrast, SNOWA records the lowest rank among customers from Tehran, Iran.

Finally, Salminen et al. (2019) conducted a study to classify the different content formats and validate the channels within companies. To carry out their research, the authors applied three fuzzy machine learning techniques: random forest, K-nearest neighbor, and neural network. The results suggest the YouTube platform as the main tool to optimize the performance of content techniques.

To make an analysis of different marketing campaigns, the approach proposed by Rancati & Gordini (2014) was used because it is the best option to characterize the performance of content marketing in companies, and because its indicators have been used in various publications. Likewise, a fuzzy logic technique will be used, particularly the OWA operator, as it is an ideal technique to classify and order campaigns according to their content capacity.

Among the main aggregator operators is the OWA operator proposed by Yager (1988), which is an operator that considers a reordering process based on two different vectors: the arguments and the weights. With this, it can be possible to generate different scenarios that are not as easy to consider just based on the average (Beliakov et al., 2007; Blanco-Mesa et al., 2019; Yager & Kacprzyk, 2012). The formulation of the OWA operator is as follows.

Definition 13.1 An OWA operator of dimension n is a mapping $OWA: R^n \rightarrow R$ with an associated weight vector W of dimension n such that $\sum_{j=1}^{n} w_j = 1$ and $w_j \in [0,1]$, according to the following formula:

$$OWA(a_1, a_2, \ldots, a_n) = \sum_{j=1}^{n} w_j b_j, \tag{13.1}$$

where b_j is the jth largest element of the collection a_i. It is important to note that based on the reordering process, it is possible to have the $OWA_{\min}$ where the higher argument has the lower weight and the $OWA_{\max}$ where the higher argument has the higher weight.

13.3 MEASUREMENT OF THE CONTENT MARKETING PERFORMANCE OF A COMPANY IN CULIACÁN WITH THE OPERATOR OWA

To measure the content marketing performance of a company in Culiacán that markets mobile phone accessories through the proposed methodology, the following steps must be followed.

Step 1. Since the objective is to measure personalized campaigns created based on content marketing, this will be the main element and is composed of four criteria and different indicators that are obtained from the literature (see Table 13.1).

Table 13.1 Indicators to measure the performance of a company's content marketing

Identifier	*Description*	*Category*	*H or L*
CON	Consumption	Criterion	
TNV	Total number of views	Indicator	H
TNU	Total number of visitors	Indicator	H
TNA	Total number of actions/clicks per content	Indicator	H
VIT	Viewing time	Indicator	H
NVV	Number of views per visitor	Indicator	H
VTV	View time per visitor	Indicator	H
EXC	Exchange	Criterion	
TNL	Total number of likes	Indicator	H
TNS	Total number of shares	Indicator	H
TNM	Total number of messages	Indicator	H
TNC	Total number of comments	Indicator	H
TNI	Total number of inbound links	Indicator	H
LEA	Leads	Criterion	
NPC	Number of people who completed the form or action	Indicator	H
TOC	Total number of comments on the site	Indicator	H
TNS	Total number of subscribers	Indicator	H
COR	Conversion rate	Indicator	H
SAL	Sales	Criterion	
TOS	Total number of sales	Indicator	H
CRR	Customer retention rate	Indicator	H
CLV	Customer lifetime value	Indicator	H
PPI	Percentage of profit on investment	Indicator	H

Step 2. Based on Table 13.1, information was obtained from each of the five personalized campaigns created by the mobile phone accessories company, and the last column (minimum) is the least that the owner wants to obtain for each indicator in any campaign (see Tables 13.2 and 13.3). Finally, it is important to note that all indicators are maximizing, that is, the higher the better. This information will be used for the variable normalization process with the following formula: $9\times\left(\dfrac{\text{score}-\text{score}_{Mn}}{\text{score}_{Mx}-\text{score}_{Mn}}\right)+1$, where score is the specific value for the State, score_{Mn} is the minimum, and score_{Mx} is the maximum value of the indicator, being 10 the highest score and 1 the lowest one, and the results can be seen in Table 13.4.

Step 3. With the information provided in Table 13.4, the use of the average of each criterion is obtained (see Table 13.5).

Table 13.2 Personalized campaigns created by the mobile phone accessories company

Identifier	*Campaigns*
A_1	Instagram messages 01
A_2	Instagram messages 02
A_3	Instagram messages 03
A_4	Instagram messages 04
A_5	Instagram messages 05

Table 13.3 Results of each identifier for each campaign

	A_1	A_2	A_3	A_4	A_5	*Minimum*
CON						
TNV	35,872.00	66,031.00	48,072.00	41,922.00	47,808.00	35,000.00
TNU	120,040.00	112,362.00	120,627.00	112,535.00	111,924.00	90,000.00
TNA	188.00	159.00	105.00	111.00	152.00	80.00
VIT	0.12	0.10	0.15	0.08	0.11	0.05
NVV	1.94	2.34	1.86	2.47	3.12	1.50
VTV	0.09	0.08	0.05	0.03	0.07	0.01
EXC						
TNL	135.00	128.00	133.00	144.00	114.00	100.00
TNS	54.00	65.00	54.00	47.00	55.00	40.00
TNM	148.00	149.00	155.00	111.00	152.00	100.00
TNC	36.00	31.00	33.00	31.00	35.00	25.00
TNI	4,180.00	5,234.00	4,188.00	6,432.00	5,461.00	3,500.00
LEA						
NPC	121.00	150.00	155.00	141.00	152.00	90.00
TOC	14.00	20.00	18.00	15.00	18.00	8.00
TNS	93.00	85.00	75.00	77.00	80.00	60.00
COR	0.27	0.19	0.20	0.26	0.22	0.10
SAL						
TOS	26.00	23.00	29.00	27.00	25.00	15.00
CRR	0.53	0.38	0.45	0.55	0.43	0.25
CLV	227.00	222.00	270.00	268.00	233.00	150.00
PPI	369.00	364.00	359.00	422.00	359.00	300.00

Step 4. The weights for each criterion must be defined. The weights were obtained from a digital marketing expert. This person has more than 15 years of experience in the area, and the weighting vector is $w = 0.33, 0.19, 0.28, 0.20$.

Table 13.4 Normalized value for each indicator

Identifier	A_1	A_2	A_3	A_4	A_5
CON					
TNV	1.25	10.00	4.79	3.01	4.71
TNU	9.83	7.57	10.00	7.62	7.44
TNA	10.00	7.58	3.08	3.58	7.00
VIT	7.30	5.50	10.00	3.70	6.40
NVV	3.44	5.67	3.00	6.39	10.00
VTV	10.00	10.00	10.00	10.00	10.00
EXC					
TNL	8.16	6.73	7.75	10.00	3.86
TNS	6.04	10.00	6.04	3.52	6.40
TNM	8.85	9.02	10.00	2.80	9.51
TNC	10.00	5.91	7.55	5.91	9.18
TNI	3.09	6.32	3.11	10.00	7.02
LEA					
NPC	5.29	9.31	10.00	8.06	9.58
TOC	5.50	10.00	8.50	6.25	8.50
TNS	10.00	7.82	5.09	5.64	6.45
COR	10.00	5.79	6.11	9.77	7.39
SAL					
TOS	8.07	6.14	10.00	8.71	7.43
CRR	9.40	4.90	7.00	10.00	6.40
CLV	6.78	6.40	10.00	9.85	7.23
PPI	6.09	5.72	5.35	10.00	5.35

Step 5. With the information of Table 13.5 and the weighting vector, the ranking of the best campaign will be obtained using different aggregation operators. In this case, it is the average, weighted average (WA), OWA_{min}, and OWA_{max}. The results are presented in Table 13.6.

Step 6. With the information provided in Table 13.6, it is possible to generate different rankings depending on the aggregation operator that is considered (see Table 13.7).

Step 7. In this step, an analysis of the results is done. With the information provided in Table 13.7, it is possible to visualize that the ranking is not always the same and that many changes can be seen, even in the OWA_{min}, where A_4 is the first campaign, even when it is the last one in all other aggregation operators. This is interesting because other techniques can be used to obtain the best campaign, even when A_1 is two times the best campaign; in the other situations, it is in the third and fourth positions. Because of that, to determine which one is the best campaign to maintain, value will be given to each position.

Table 13.5 Average of each criterion

Identifier	A_1	A_2	A_3	A_4	A_5
CON	6.9708	7.7202	6.8124	5.7170	7.5929
EXC	7.2282	7.5954	6.8895	6.4458	7.1948
LEA	7.6981	8.2292	7.4244	7.4299	7.9818
SAL	7.5841	5.7910	8.0881	9.6411	6.6015

Table 13.6 Campaign score using different aggregation operators

Identifier	A_1	A_2	A_3	A_4	A_5
Average	7.3703	7.3340	7.3036	7.3085	7.3428
WA	7.3460	7.4532	7.2535	7.1199	7.4279
OWA_{min}	7.4340	7.4967	7.4202	7.6373	7.4533
OWA_{max}	7.3037	7.1454	7.1988	7.0092	7.2282

Table 13.7 Marketing campaign ranking using different aggregation operators

Identifier	*Ranking*
Average	$A_1 > A_5 > A_2 > A_4 > A_3$
WA	$A_2 > A_5 > A_1 > A_3 > A_4$
OWA_{min}	$A_4 > A_2 > A_5 > A_1 > A_3$
OWA_{max}	$A_1 > A_5 > A_3 > A_2 > A_4$

If the campaign is the first in the ranking, it will have one point, the second will have two points, and so on. With that operation, the results are $A_1 = 9$, $A_2 = 10$, $A_3 = 17$, $A_4 = 15$, and $A_5 = 9$. In the end, the best alternative can be A_1 and A_5.

As can be seen from the results, the relative importance of each criterion is essential to obtain the best alternative, in the sense that each company can have a different perspective about what is more critical for them. This can limit the methodology because the results cannot be applied to all companies, or the same campaign can be ranked number one if the weighting vector changes. Finally, this limitation is also something important in the decision-making process. The way that one's results are not always better than those of others reflects the importance of the specific situation that each company is having. In that sense, considering the analysis's relative importance improves the problem's visualization. It can help to understand the same better, becoming an important element in making the final decision.

13.4 CONCLUSIONS

This chapter presents a measurement of the content marketing performance of a company that markets mobile phone accessories through the OWA and its extensions. A fuzzy system tool was used to compare different levels of content marketing based on different aggregation operators. With the obtained results, the relative weight of each criterion is essential for determining the best campaign, and considering that not all marketing campaigns can have the same purpose, it is necessary to decide the importance of each criterion before evaluation.

According to the information in Section 13.4, there is an interesting case of campaign A_4, which can be considered the worst campaign according to WA and OWA_{max} but is the best campaign based on OWA_{min}. These results show that the ranking can change drastically based on the weights, and with that idea, the decision-maker must have a better idea of how the weights must be assigned and how little change can drastically change the order of the campaigns. Also, it is possible to assume that A_1 is the best campaign, being ranked the number one twice with the Average and OWA_{max} operators.

In the future research, the use of different aggregation operators that can help to solve this complex problem can be used, such as the Heavy OWA (Olazabal-Lugo et al., 2019; Yager, 2002) that uses a weighting vector with a sum different than one or the prioritized OWA (Espinoza-Audelo et al., 2021; Yager, 2008) that can help to unify the information of other decision-makers or experts in the problem. In addition, it is necessary to identify which indicators can be added or should be removed from the formulation to increase confidence in the instrument.

REFERENCES

Bala, M., & Verma, D. (2018). A critical review of digital marketing. *International Journal of Management, IT & Engineering*, *8*(10), 321–339.

Baltes, L. P. (2015). Content marketing-the fundamental tool of digital marketing. *Bulletin of the Transilvania University of Brasov. Series V: Economic Sciences*, *8*, 111–118.

Bangsawan, S., Mahrinasari, M., & Ahadiat, A. (2021). Digital content marketing strategy in increasing customer engagement in Covid-19 situation. *International Journal of Pharmaceutical Research*. doi: https://doi.org/10.31838/ijpr/2021.13.01.684

Beliakov, G., Pradera, A., & Calvo, T. (2007). *Aggregation functions: A guide for practitioners* (Vol. 221). Heidelberg, Germany: Springer.

Blanco-Mesa, F., León-Castro, E., & Merigó, J. M. (2019). A bibliometric analysis of aggregation operators. *Applied Soft Computing*, *81*, 105488.

Du Plessis, C. (2017). The role of content marketing in social media content communities. *South African Journal of Information Management*, *19*(1), 1–7.

Espinoza-Audelo, L. F., Leon-Castro, E., Mellado-Cid, C., Merigo, J. M., & Blanco-Mesa, F. (2021). Uncertain induced prioritized aggregation operators in the analysis of the imports and exports. *Journal of Multiple-Valued Logic & Soft Computing*, *36*(6), 543–568.

Ho, J., Pang, C., & Choy, C. (2020). Content marketing capability building: A conceptual framework. *Journal of Research in Interactive Marketing*, *14*(1), 133–151.

Hollebeek, L. D., & Macky, K. (2019). Digital content marketing's role in fostering consumer engagement, trust, and value: Framework, fundamental propositions, and implications. *Journal of Interactive Marketing*, *45*(1), 27–41.

Koob, C. (2021). Determinants of content marketing effectiveness: Conceptual framework and empirical findings from a managerial perspective. *PLoS One*, *16*(4), e0249457.

Mansour, D., & Barandas, H. (2017). High-tech entrepreneurial content marketing for business model innovation: A conceptual framework. *Journal of Research in Interactive Marketing*, *11*(3), 296–311.

Müller, J., & Christandl, F. (2019). Content is king–But who is the king of kings? The effect of content marketing, sponsored content & user-generated content on brand responses. *Computers in Human Behavior*, *96*, 46–55.

Naseri, Z., Noroozi Chakoli, A., & Malekolkalami, M. (2023). Evaluating and ranking the digital content generation components for marketing the libraries and information centres' goods and services using fuzzy TOPSIS technique. *Journal of Information Science*, *49*(1), 261–282.

Naseri, Z., & Noruzi, A. (2018). Content marketing process model: A meta-synthesis of the literature. *Webology*, *15*(1), 8–18.

Olazabal-Lugo, M., Leon-Castro, E., Espinoza-Audelo, L. F., Maria Merigo, J., & Gil Lafuente, A. M. (2019). Forgotten effects and heavy moving averages in exchange rate forecasting. *Economic Computation and Economic Cybernetics Studies and Research*, *53*(4), 79–96.

Rancati, E., & Gordini, N. (2014). Content marketing metrics: Theoretical aspects and empirical evidence. *European Scientific Journal*, *10*(34), 92–104.

Salminen, J., Yoganathan, V., Corporan, J., Jansen, B. J., & Jung, S.-G. (2019). Machine learning approach to auto-tagging online content for content marketing efficiency: A comparative analysis between methods and content type. *Journal of Business Research*, *101*, 203–217.

Taheri, A., & Bashekouh Ajirlo, M. (2023). Development of an artificial intelligence model based on fuzzy logic to content marketing capability analysis for buying TV brands in Tehran on Telegram. *Advances in Cognitive Science*, *25*(3), 108–127.

Taiminen, K., & Ranaweera, C. (2019). Fostering brand engagement and value-laden trusted B2B relationships through digital content marketing: The role of brand's helpfulness. *European Journal of Marketing*, *53*(9), 1759–1781.

Wall, A., & Spinuzzi, C. (2018). The art of selling-without-selling: Understanding the genre ecologies of content marketing. *Technical Communication Quarterly*, *27*(2), 137–160.

Yager, R. R. (1988). On ordered weighted averaging aggregation operators in multicriteria decisionmaking. *IEEE Transactions on Systems, Man, and Cybernetics*, *18*(1), 183–190.

Yager, R. R. (2002). Heavy OWA operators. *Fuzzy Optimization and Decision Making*, *1*(4), 379–397.
Yager, R. R. (2008). Prioritized aggregation operators. *International Journal of Approximate Reasoning*, *48*(1), 263–274.
Yager, R. R., & Kacprzyk, J. (2012). *The ordered weighted averaging operators: Theory and applications*. New York: Springer Science & Business Media.

Index

Pages in *italics* refer to figures and pages in **bold** refer to tables.

For Product Safety Concerns and Information please contact our EU representative GPSR@taylorandfrancis.com Taylor & Francis Verlag GmbH, Kaufingerstraße 24, 80331 München, Germany